NEIL F. COMINS

THE TRAVELER'S GUIDE TO SPACE

———

For One-Way Settlers
and Round-Trip Tourists

COLUMBIA UNIVERSITY PRESS
NEW YORK

Columbia University Press
Publishers Since 1893
New York   Chichester, West Sussex
cup.columbia.edu
Copyright © 2017 Neil F. Comins
All rights reserved

Library of Congress Cataloging-in-Publication Data

Names: Comins, Neil F., 1951–
Title: The traveler's guide to space : for one-way settlers and round-trip
tourists / Neil F. Comins.
Description: New York : Columbia University Press, [2017] | Includes
bibliographical references and index.
Identifiers: LCCN 2016026420 (print) | ISBN 9780231177542 (cloth : alk. paper) |
ISBN 9780231542890 (electronic)
Subjects: LCSH: Space flight—Physiological effect. | Manned space flight.
Classification: LCC RC1150 .C66 2017 (print) | LCC RC1150 (ebook) |
DDC 616.9/80214—dc23
LC record available at https://lccn.loc.gov/2016026420

Columbia University Press books are printed on permanent
and durable acid-free paper.

Printed in the United States of America

Cover Image: ©Michael Seeley / Demotix / Corbis

CONTENTS

Introduction   *vii*

PART I: PREPARING FOR SPACE

▬

1. SCIENCE AND THE SOLAR SYSTEM OVER EASY   3

2. BRIEF DESCRIPTIONS OF JOURNEYS THROUGH SPACE   26

3. PREPARING FOR YOUR TRIP   42

4. TRAINING FOR SPACE TRAVEL   48

PART II: ADJUSTING TO SPACE

▬

5. LAUNCH!   63

6. ADJUSTMENTS DURING THE FIRST FEW DAYS   68

7. LONG-TERM PHYSICAL ADJUSTMENTS TO SPACE   74

8. GETTING ALONG IN SPACE: PSYCHOLOGICAL AND
SOCIOLOGICAL ASPECTS OF SPACE TRAVEL   107

**PART III: MAKING THE MOST
OF EXPERIENCES IN SPACE**

▬

9. EXPERIENCES BY DESTINATION   145

PART IV: HOME! SWEET? HOME?

▬

10. EMIGRATING TO MARS OR
RETURNING TO EARTH   233

Appendix: Powers of Ten   253
Notes   255
Bibliography   259
Index   275

INTRODUCTION

O rganized travel for the wealthy and powerful began thousands of years ago in countries around the world including Egypt, Greece, and China. With the formation of the "middle class" and the development of the printing press in the fifteenth century C.E., the ability of growing numbers of people to plan and carry out tourist travels increased dramatically, until today there are over a billion people taking tourist vacations annually.

The first self-funded space tourism trip was made in 2001 by U.S. entrepreneur Dennis Tito, who went to the *International Space Station* for an eight-day vacation. As of June 2016, seven paying tourists have flown in space, including U.S. billionaire Charles Simonyi, who has gone twice. All of them have stayed in orbit around the Earth, visiting the *International Space Station*.

Several countries and international consortia, such as the European Space Agency, have been involved in developing and implementing space exploration technology for decades, and now private corporations are also actively developing space travel hardware. Some of the corporate leaders are among the visionaries hoping to be involved in colonizing the Moon and Mars and in developing space tourism opportunities in the very near future.

Virtually every aspect of life changes when one gets into space, from ordinary activities such as moving and eating to the internal workings of our bodies to the way we have sex. Images of other worlds, including the Moon, Mars and its two moons, asteroids (rock and metal space debris), and comets (rock and ice space debris) taken by satellites visiting these bodies reveal that each one has unique features, many of which are also unlike anything seen on Earth. Therefore, virtually everything visitors to these worlds would see and do will be different in many ways from analogous experiences on Earth. I have written *The Traveler's Guide to Space* with the hope of providing people interested in learning more about our local region of the cosmos with information about the opportunities and challenges that space travelers can expect to experience.

The book uses virtually no mathematics (other than the occasional shorthand "powers of ten" and the equation $F = ma$, where F is force, m is mass, and a is acceleration), and I explain all the science necessary to understand the what and why of life in space. Material in the book pertaining to various challenges to life in space was drawn from research I did for a previous book, *The Hazards of Space Travel*. I wish to thank my agent, Louise Ketz, for handling the business matters related to the book; my editor, Patrick Fitzgerald, for refining the work; Ryan Groendyk, for helping organize figures and other materials included in the book; Professor Andrew West for his thoughtful edits; and my textbook publisher, W. H. Freeman & Co., for permission to use several figures from books I have written for them. I especially want to thank my wife, Sue, for her patience as I wrote this book and for her thoughtful review of the manuscript with her eye as an English professor.

THE TRAVELER'S GUIDE TO SPACE

———

Part I

PREPARING FOR SPACE

1

SCIENCE AND THE SOLAR SYSTEM OVER EASY

—

Human spaceflight began on April 12, 1961, when Soviet cosmonaut Yuri Gagarin was launched from the Baikonur Cosmodrome in what is now Kazakhstan. He rode a Vostok-K rocket eastward to an altitude of approximately 325 kilometers (km) (203 miles). Over Africa, after the rocket had nearly circumnavigated the globe in space, retro-rockets were fired, slowing his space capsule and allowing it to descend back toward the Earth. At an altitude of 7,000 meters (m) (23,000 feet), the door of the space capsule was blown off (intentionally) and Yuri was ejected into the air. He parachuted to Earth, some 280 km (175 mi) from his intended landing site, and from there phoned for a ride home.

Human spaceflight has moved forward in fits and starts since then. The 1960s and '70s were witness to the Soviet–U.S. space race, culminating in humans visiting the Moon (only our moon is always capitalized). Earth-orbiting space stations, including the Soviet *Salyut* series, the Soviet/Russian *Mir* space station, the U.S. *Skylab*, and the Chinese *Tiangong 1*, have all come and gone. Today the *International Space Station* and the Chinese *Tiangong 2* orbit in majestic isolation. The International Space Station provides countries, companies, and organizations the opportunity to send not only astronauts and scientists but also nonprofessional travelers willing and able to pay for such trips into space. To date, between 500 and 1,000 people have traveled into space. (Classified military missions limit the precision of this number.) Numerous commercial space ventures are under development, bringing with them the promise of many more human space travel opportunities in the very near future.

While prices for trips into space by nonprofessional travelers have historically been in the tens of millions of U.S. dollars, the new generation of companies involved in all aspects of spaceflight has the potential to drive the prices down. The cost of a seat on one of the short tourist flights into space presently under development is expected to be only a few hundred thousand dollars. This means that the opportunity to go into space will soon be available to millions of people.

In order to make the most of a journey into space, it is very important to understand the basic scientific and medical issues you are likely to encounter out there. In explaining the various aspects of the human experience in space, *The Traveler's Guide to Space* therefore contains a fair amount of science. Making these concepts meaningful to you is one of the goals of this book. One major problem laypeople confront is that scientists imbue certain everyday words with very technical meanings. So defining the meanings of words in scientific contexts is done as needed throughout the book.

Let's consider two examples of how scientific words can be interpreted differently. As you probably know, tiny electrons orbit the much larger nuclei of atoms. Hearing that, most people visualize electrons as tiny planets and the nuclei as analogous to tiny suns. Indeed, in the early twentieth century, there were even stories about microscopic life on atoms (see, e.g., *Girl in the Golden Atom* by Ray Cummings, published in 1922). In reality, however, none of the atomic particles—protons, neutrons, or electrons—is at all solid. Rather, they each have properties of both solids and waves. That means, among other things, that an electron cannot be envisioned as a planet orbiting a star—the electron's wave properties distribute it all the way around the atom's nucleus as it orbits, rather than it moving as a pointlike object. While this complicates "visualizing" electrons, thinking of them as having both particle and wave properties enables scientists to explain their behavior in exquisite detail.

The first sentence of this chapter is another excellent example of how important it is to understand the meaning of scientific words. There I wrote, "Human spaceflight . . ." without defining "space." The intuitive definition most of us have is that space is that region above our atmosphere. The problem is that the atmosphere, unlike the surface of the water in a swimming pool, lacks a definitive upper boundary. As you descend into the atmosphere from far away, you encounter increasingly dense air, unlike what happens when you jump into a swimming pool. You know for sure when you hit the well-defined surface of the water. Without a solid top, the atmosphere alone can't be used to define the beginning of "space" above the Earth. We need to approach the meaning of the term from a slightly different angle.

The nature of a gas explains why the Earth's atmosphere lacks a well-defined surface. Liquids are weakly bound groups of atoms and molecules (bound sets of atoms); gases are atoms and molecules that are unable to bind together. Unless held near each other by some external means, they drift apart. For example, a balloon holds the gas atoms or molecules in it near each other. Once it is pierced or opened, the atoms or molecules inside it spread farther apart.

In rough analogy to how a balloon holds gases close together, the Earth's gravity holds the gases of the atmosphere close to the Earth, preventing most of them from drifting into space. Granted, the uppermost gases, heated by energy from the Sun, are continually escaping Earth's gravity and drifting into interplanetary space. They do this because the more anything is heated, the faster the particles in it move. Gases moving fast enough can escape the Earth's gravitational attraction. As a result, some of the Earth's atmosphere continually drifts out to the distance of the Moon and beyond, never to return here. Complicating matters is the fact that depending on the temperature of the air, whole regions of the atmosphere sometimes expand upward (when heated) and sometimes sink down (when cooled). Therefore, the height of the atmosphere is continually changing with the weather, the amount of energy from the Sun, and the day-and-night cycle.

Most space enthusiasts define "space" as the altitude of the Karman line and above. This definition evolved from the reasoning of Theodore von Karman (1881–1963), a Hungarian American scientist. Karman asked: How dense does the air have to be in order for the wings of an aircraft to generate enough lift to keep the plane up? After all, the higher you go, the less dense the air, and hence the lower the amount of lift it provides at any speed. He found that at an altitude of about 100 km (62 mi) above the Earth's surface, the air is so thin that in order to keep itself up, a plane would have to travel so fast that it would put itself into orbit. In other words, at that altitude the plane would be going so fast that you could turn off its engines and it would continue to orbit around the Earth. This altitude is now called the Karman line.

Objects orbiting without the use of engines use their speed to prevent themselves from falling to Earth. For example, the *International Space Station*[1] is continually being pulled earthward by gravity, but its motion parallel to the Earth is fast enough that as it falls, it continually avoids hitting the planet. For various political and technical reasons, some organizations and countries define "space" as starting at different altitudes, but for our purpose, it begins at the Karman line.

■ ■ ■

Having established where space begins, we now consider the destinations in it that space travelers can expect to visit in the near future. It will help to have a cosmic perspective in this matter. We live in the solar system, defined as the Sun and everything that orbits it. All of the objects orbiting the Sun are divided into four classes: planets, moons, dwarf planets, and small solar system bodies, defined by the International Astronomical Union in 2006.

Astronomers in the International Astronomical Union define planets in the solar system[2] as having two properties. First, they must have enough mass that their own gravity causes them to be roughly spherical in shape. Indeed, Earth, with its mountains and valleys, is not perfectly spherical. Also, our planet's rotation causes it to be wider at the equator than it is between the North and South poles: Earth is about[3] 42.8 km (26.6 mi) wider at the equator than it is from pole to pole. These small variations from spherical don't affect the definition of a planet. Second, planets must have enough mass that their gravitational attraction is strong enough either to pull nearby space debris onto them or to fling it far away. Put another way, planets can clear their orbits of debris. This does not include their moons, smaller bodies that are gravitationally bound in orbit to their respective planets.

The present list of planets in our solar system is as follows. Going outward from the Sun, the eight planets are Mercury, Venus, Earth, Mars, Jupiter, Saturn, Uranus, and Neptune. There are at least 173 moons orbiting the various planets; only Mercury and Venus lack moons. In 2015, astronomers discovered that the paths of some objects far beyond Neptune's orbit are being affected by the gravitational pull of an as yet unseen body. Since calculations suggest that this is a planet-sized object, if and when it is discovered, it will be catalogued as the ninth planet in our solar system.

All the planets orbit the Sun in essentially the same plane as we do, and they are going around it in the same direction as the Earth travels. The plane defined by the Earth's orbit around the Sun is called the ecliptic.

Pluto lost planet status in the new classification scheme because its mass is much lower than the masses of the other planets. This really does matter. The mass of any object is a measure of how much matter it contains or, putting the same concept in different terms, how many particles of each type of element it has.

Pluto is big and massive enough to be spherical, but it lacks sufficient mass to clear its neighborhood of smaller debris. It is therefore classified as a dwarf planet. Five bodies in the solar system are identified as dwarf planets: Pluto, Ceres, Haumea, Makemake (pronounced mah-kay-mah-kay), and Eris. Pluto, Haumea, Makemake,[4] and Eris are known to have moons. Astronomers are evaluating the properties of other bodies in our solar system to see if they meet the dwarf planet criteria.

Small solar system bodies are the final set of objects orbiting the Sun. These are essentially pieces of debris composed primarily of rock and ice or rock and metal, held together mainly by their chemical bonds rather than by their gravitational self-attraction, as are planets and dwarf planets.

Both dwarf planets and small solar system bodies are mixes of objects formerly classified as asteroids, meteoroids, and comets. Asteroids and meteoroids comprise all the space debris in the solar system composed primarily of rock and metal. The larger such bodies were called asteroids, although there was no official boundary in size between them and meteoroids. Common usage is that meteoroids are smaller than about a meter (1 yard) across. Several asteroids are known to have moons. Complicating matters slightly is the fact that the moons of asteroids are also asteroids. The smaller of each pair of orbiting asteroids is considered the moon of the larger one.

ASTEROIDS

Every asteroid is unique in appearance, from spherical to peanut shaped to rubber duck shaped and more. They were formed chaotically early in the life of the solar system. Most of the asteroids are located in relatively circular orbits in the region of the solar system between Mars and Jupiter, called the Asteroid Belt. Other asteroids are in more elliptical (oval-shaped) orbits that cross the paths of one or more planets. Asteroids are being discovered nearly every day now; over 450,000 have been catalogued, with sightings of several hundred thousand more awaiting final confirmation. Despite big-budget Hollywood movies that show us otherwise, visiting an asteroid will not require your ship's captain to dodge numerous other asteroids on your way. While there are lots of asteroids orbiting the Sun, they are typically 1.6 million km (1 million mi) apart.

Near Earth Objects orbit at nearly the same distance from the Sun as we do. Over 12,000 Near Earth Objects are known. Four groups of them are asteroids that are candidates for visits in the near future: the Amors, Apollos, Atens, and Atira. The Amors have orbits between Mars and Earth. They always remain farther out from the Sun than the Earth, but they do pass close to us. NASA landed a satellite on the Amor asteroid Eros in 2001 (figure 1.1). Conversely, the Atira or Interior Earth Objects are asteroids that orbit inside Earth's orbit, but never come out as far as we are.

The Apollos and Atens are Earth-crossing asteroids. Their orbits are so elliptical that they are sometimes closer to the Sun than Earth and sometimes farther away. Some of them pass extremely close to us, and some have smashed into Earth, such as the ones that caused the Cretaceous-Tertiary (K-T) extinction 65 million years ago and possibly the even more horrific Permian-Triassic extinction (also called The Great Dying) that occurred about 250 million years ago.

SCIENCE AND SCIENCE FICTION

Asteroids have played key roles in dozens of science fiction stories dating back to the work of Jules Verne in the late nineteenth century. The Star Wars franchise used them in several episodes. In *The Empire Strikes Back*, Han Solo hides his *Millennium Falcon* in an asteroid that is part of a closely packed belt, consistent with the common image of the Asteroid Belt in our solar system. The problem is that if asteroids were as close together as they are shown to be in such movies, their mutual gravitational attraction would cause them to collide with one another within years of their being created. They would, in all likelihood, form a single large body. The reason this doesn't happen in real life is that asteroids in the Asteroid Belt, as well as the Near Earth Objects, are separated from one another by at least 1.6 million km (1 million mi, four times the distance from the Earth to the Moon) and possibly several times greater distance, depending on whose calculations you believe. At those distances, their relatively low masses and relatively high velocities do not allow them to pull together.

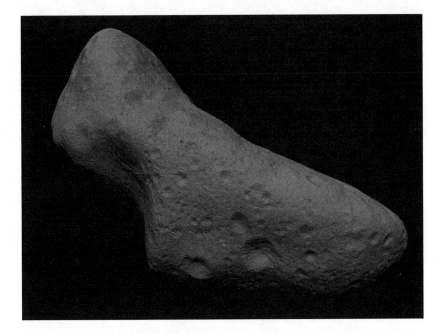

FIGURE 1.1

(a) Discovered in 1898, the Amor asteroid Eros has been as close as 26.7 million km (16.6 million mi) from Earth. Its longest dimension is about 34 km (21 mi) across.

NEAR Project, NLR, JHUAPL, Goddard SVS, NASA

FIGURE 1.1 *(continued)*

(b) Close-up of the surface of Eros, taken by the lander NEAR Shoemaker.

NEAR Project, NLR, JHUAPL, Goddard SVS, NASA

The distinction between the Apollo and Aten asteroids is that the Apollo asteroids take more than one Earth year to orbit the Sun, while the Atens orbit in less than a year. As a result, it takes each Apollo asteroid more than a year to cross our path (inward and then out again) once, while the Atens cross our orbit more than once a year. NASA has a catalog containing over 1,300 candidate Near Earth Objects for humans to visit in the near future, and that number is growing rapidly. Figure 1.2 summarizes the orbits of Near Earth Objects.

The final class of asteroids relevant to space travel in the near future are those locked in the same orbit as the Earth. These are called Trojan asteroids, named after participants in the Trojan War. Mars, Jupiter, Uranus, and Neptune also have Trojans, and the other planets may have them as well.

It may seem that the Trojans, orbiting in the same orbits as the planets, violate the definition of planets as clearing their orbits of other debris, but

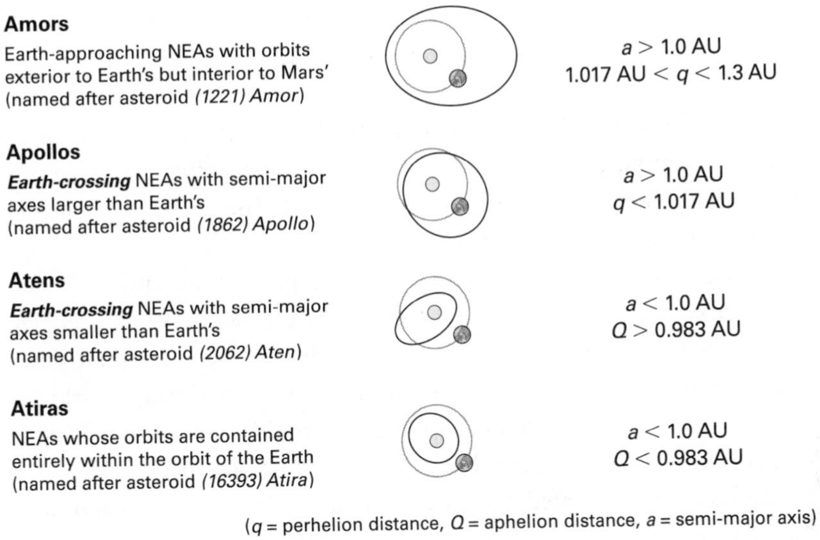

Amors

Earth-approaching NEAs with orbits exterior to Earth's but interior to Mars' (named after asteroid *(1221) Amor*)

$a > 1.0$ AU
1.017 AU $< q < 1.3$ AU

Apollos

Earth-crossing NEAs with semi-major axes larger than Earth's (named after asteroid *(1862) Apollo*)

$a > 1.0$ AU
$q < 1.017$ AU

Atens

Earth-crossing NEAs with semi-major axes smaller than Earth's (named after asteroid *(2062) Aten*)

$a < 1.0$ AU
$Q > 0.983$ AU

Atiras

NEAs whose orbits are contained entirely within the orbit of the Earth (named after asteroid *(16393) Atira*)

$a < 1.0$ AU
$Q < 0.983$ AU

(q = perhelion distance, Q = aphelion distance, a = semi-major axis)

FIGURE 1.2

The different types of Near Earth asteroids. Perihelion is the point of closest approach of a body to the Sun. Aphelion is the farthest point from the Sun in a body's orbit. The semimajor axis is half the distance from aphelion to perihelion.

NASA

there is a good reason this rule does not apply to these asteroids. There are two places in each planet's orbit where the combined gravitational attractions of the planet and the Sun hold asteroids securely in place. These locations, called the L4 and L5 Lagrange points, are 60° ahead of the planet in its orbit and 60° behind the planet in its orbit. (As you surmised from the numbers, there are other Lagrange points, L1, L2, and L3, all of which are points where any piece of space debris will be pulled unceremoniously away by the gravitational attractions of its planet and the Sun.) Because asteroids at our L4 and L5 locations never orbit close to the Earth, the Earth's gravity cannot cause them to either crash into us or be flung far away. At present, Earth is known to have one Trojan asteroid. Astronomers are looking for others.

Asteroids have a variety of compositions worth knowing about. The smaller ones are essentially giant boulders in space. The much larger ones, however, formed from myriad collisions of smaller bodies. These impacts heated the growing asteroids, making them molten and enabling the dense metals in them, like iron and nickel, to sink to their cores. This forced the lighter rock they contained to float on the cores and create their outer layers. Then they cooled and solidified, locking into place this structure of metal deep inside and rock outside. The Earth has a similar structure of a metallic core surrounded mostly by rock.

The story doesn't always end there. Some large asteroids endured powerful impacts after they solidified. Depending on their sizes and relative speeds, some asteroids involved in these collisions were completely destroyed, with their remnant metal and rock shards drifting in orbit as smaller asteroids and meteoroids. Other asteroids involved in collisions ejected some of their content but otherwise remained intact. These two collision processes created many of the smaller asteroids and meteoroids. The reason this may be important for space travelers is that some remnants of larger asteroids, such as their metal cores, have significant value, both as collectibles and for commercial purposes on Earth.

The surfaces of all dwarf planets, small solar system bodies, and moons evolve over time. Since they solidified, their rocky surfaces have been continually pounded and pulverized by impacts from countless meteoroids, as well as from high-speed particles ejected from the Sun, called the solar wind, and from particles originating outside the solar system called cosmic rays. As a result, their surfaces are for the most part powdery, with craters (see figure 1.1a) and rocks and boulders sticking up through the powdered surface, called the regolith (figure 1.3).

COMETS

The other space debris orbiting the Sun are the comets, some of which are also suitable candidates for space travel in the near future. Comets are composed primarily of rock and ice. It is important to note that in astronomy, the word "ice" covers frozen water, frozen carbon dioxide, frozen carbon monoxide, frozen methane, and frozen ammonia.

The evolution of comets is interesting and relevant to space travel to them. Comets came into existence when the solar system was first form-

FIGURE 1.3

Apollo 11 footprint in the regolith of the Moon.

NASA

ing from the collisions of myriad dust and gas particles. Comet formation took place far enough from the young Sun that it was sufficiently cold for the dust and gas particles to bond together. As they did so, larger and larger clumps of ice and rock formed. The bodies of rock and ice that exist today in orbit around the Sun are called comet nuclei.

Unlike all the other objects discussed so far, most comet nuclei are not orbiting near the plane of the ecliptic but in two regions far out beyond the planets. The Kuiper belt of comets and other space debris, named after the Dutch American astronomer Gerard Kuiper (1905–1973), who first proposed its existence, is a bagel-shaped region of space beyond the orbit of Neptune, with the middle of the bagel, cut the long way, lying in the plane of the ecliptic. Pluto, Haumea, and Makemake also orbit in the

Kuiper belt. Much of the matter there is believed to be leftover debris from the formation of the planets.

The other major reservoir of comets is called the Oort comet cloud, after Jan Oort (1900–1992), the Dutch astronomer who first proposed the presence of these comets. The Oort cloud is a spherical distribution of comets located out beyond the Kuiper belt. This debris is believed to have been ejected from the inner solar system by the gravitational forces of the planets, especially Jupiter. Figure 1.4 shows the locations of the Kuiper belt and Oort cloud.

Over the 4.6 billion-year life of the solar system, most comet nuclei have remained in orbit beyond Neptune. However, occasionally some of these dirty icebergs in space have been nudged or pulled into orbits that take them into the inner solar system, where the planets and Sun reside. When comets come closer to the Sun than about the orbit of Neptune,

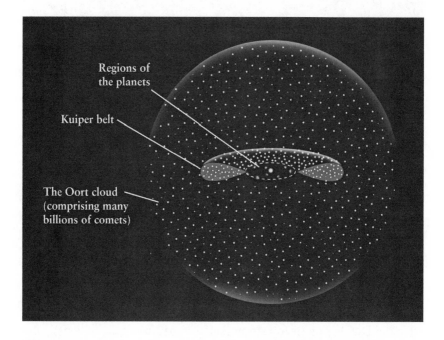

FIGURE 1.4

Locations of the Kuiper belt and the Oort comet cloud relative to the planets.

W. H. Freeman & Co.

the heat and particles streaming from the Sun cause some of the ices on them to vaporize, forming very thin atmospheres around them. These spherical atmospheres, called comas, are much less dense than the air we breathe. Because the comets have so little mass, this gas drifts into space, never to return.

Comet tails only exist when the comet nuclei are in the realm of the planets. Along with the liberated gas, pieces of dust and small pebbles freed from the ice drift into space. When these comets get close enough to the Sun, the sunlight and particles from the Sun push some of the coma's gas and debris away, creating two comet tails: one of dust particles and one of gas particles (figure 1.5). The tail of lightweight gas particles points directly away from the Sun, while the sunlight and particles from the Sun cannot push the dust away as effectively. Therefore, the dust tail points between the gas tail and the direction from which the comet came.

Different comets have different fates. Some of them come in from their distant reservoirs and fall straight into the Sun, where they are vaporized. Many more miss the Sun as they pass closest to it. These surviving comets are divided into two groups. Long-period comets are those that pass close

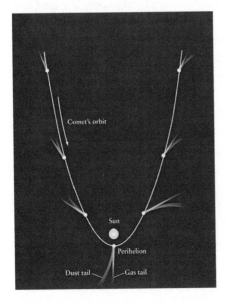

FIGURE 1.5

The tails of a comet, shown as it orbits closest to the Sun. The tail pointing directly away from the Sun is composed of gases, while the curved tail is composed of dust.

W. H. Freeman & Co.

to the Sun once and then rush back out beyond the planets, not to return for at least 200 years. Indeed, comets coming from the Oort comet cloud can have periods of over 30,000 years! Short-period comets pass near planets whose gravitational attraction causes the comet orbits to change so that they stay in the inner solar system with the planets. These comets pass close to the Sun every 200 years or less. For example, Halley's comet, which goes out to just beyond Neptune's orbit, has a 76-year orbit.

Clearly, since their comas and tails are matter that leaves a comet, never to return, comets get smaller after each close passage to the Sun. Some have layers of rocky and dusty debris that stop their dissipation. Most, however, eventually come apart, leaving rocky debris in what used to be their orbits. It is important in space travel to avoid this debris, which could collide with spacecraft.

Despite the debris they eject, travel to comets is indeed possible. The European Space Agency set the lander *Philae* onto Comet 67P/Churyumov-Gerasimenko on November 12, 2014 (figure 1.6). The lander bounced and apparently ended up in a ravine where it could not get enough sunlight to keep its batteries charged in the cold of space. Nevertheless, as it passed close to the Sun in June 2015, the lander was warmed enough to wake up and send information about the comet back to Earth.

Fittingly, comets often look like smudges on photographic images (film or charge-coupled devices [CCDs]). They are named after the people who discover them. The comet mentioned in the preceding paragraph was named after both astronomer Svetlana Ivanovna Gerasimenko, who took a photograph of it in 1969 at an observatory in southeastern Kazakh (now Kazakhstan), and astronomer Klim Ivanovych Churyumov, who discovered the comet on that photographic plate. The 67P in its name means that it is the 67th short-Period comet to be discovered.

Besides some asteroids and comets, three moons are plausible targets for space travel today: our Moon and the two moons of Mars. We will examine them in detail shortly.

Now that we have surveyed the content of our solar system, it is easy to see why human space travel will be limited to this part of the cosmos for a long time to come. The closest known star to the solar system is Proxima Centauri, located 4.0×10^{13} km (2.5×10^{13} [25 U.S. trillion, or million million] mi) from here.[5] That distance is so great that traveling at

FIGURE 1.6

Comet 67P/Churyumov-Gerasimenko imaged by the *Rosetta* spacecraft. Its longest dimension is 4 km (2.5 mi) across.

ESA/Rosetta/MPS for OSIRIS Team MPS/UPD/LAM/IAA/SSO/INTA/UPM/DASP/IDA

normal spacecraft speeds of around 64,000 km per hour (40,000 miles per hour), it would take over 70,000 years to get there from Earth. Even human travel to planets out beyond Mars or inside Earth's orbit to Venus or Mercury is not yet viable. The trips to asteroids in the Asteroid Belt or to Jupiter would take several years and are logistically very complicated. Trips by humans to Venus will likely never occur because that planet is surrounded by an atmosphere rich in sulfuric acid and so dense that at the planet's surface, the air pressure is equivalent to what you would feel 915 m (3,000 ft) underwater in Earth's oceans—a crushing experience. Finally, human travel to Mercury, the innermost planet, would also require too many years and so much new technology (to keep spacecraft that close to the Sun from overheating and also from being damaged by radiation) that we won't be going there soon.

In summary, these are several potential space travel options in the very near future:

- suborbital flights that go into space in an arc and bring you back to Earth without ever going into orbit
- orbital flights to the *International Space Station* or to commercial stations under development
- the Moon
- passing asteroids or comets
- Earth's Trojan asteroid(s)
- the moons of Mars
- Mars

We will consider all these trips further in chapter 2.

Different journeys through space take significantly different lengths of time. Your travel time will depend on several factors, including the relative distance and speed between Earth and your destination both on your trip out and on your trip home; the speed of your spacecraft; and the time you spend at your destination. Since we have limited space travel in the near future to the inner solar system, consider some of the distances involved just in our neighborhood. The greatest distance between any two points on the Earth's surface is about 2.0×10^4 km (1.25×10^4 [12.5 thousand] mi). The average distance between the Earth and the Moon is just less than 3.9×10^5 km (2.4×10^5 [240 thousand] mi). I say "average" because the distance between the Earth and the Moon varies by about 4.2×10^4 km (2.6×10^4 mi) from one new moon phase to the next. The average distance between the Earth and Sun is 150×10^6 km (93×10^6 [93 million] mi. Since Earth's orbit is elliptical, that distance varies by about 4.8×10^6 km (3.0×10^6 mi) throughout the year.

Let's consider the case of Mars, keeping in mind that the results are conceptually the same for most of the other bodies we can visit soon. Because the Earth and Mars are at very different distances from the Sun, the lengths of time it takes the two planets to orbit it are very different. As a result, sometimes they are on the same side of the Sun and sometimes they are on opposite sides of it from each other. The shortest distance between Earth and Mars is about 55×10^6 km (34×10^6 mi), while the greatest distance between them is a whopping 400×10^6 km (250×10^6 mi). The time it takes to get there from here therefore depends on the relative positions on the two worlds when you leave and the speed of your spacecraft (which determines the path you will take). Typical transit times to Mars or back range from roughly five to ten months.

■ ■ ■

Two other aspects of science are essential to understanding your experience in space. First is the nature of the particles that compose matter. This is relevant because you will be interacting not only with particles normally existing inside spacecraft but also with particles that come from outside and strike you. These represent a different mix of matter than you encounter on Earth. The second important consideration is the nature of light and related electromagnetic radiations. This is relevant because you will be exposed to a variety of radiations in space that do not normally exist on the Earth's surface. Many of these are potentially hazardous to your health, so knowing what they are and how to protect yourself are both important. We will explore matter first.

We need to understand the nature of atoms, which are individual copies of an element, like carbon or hydrogen. Basically, an atom is composed of three kinds of particles: protons, neutrons, and electrons. The protons and neutrons are bound together in a clump called the nucleus, while the electrons, which are much less massive, orbit around the nucleus. Keep in mind that the electron is not a solid particle, but rather a distribution of matter spread out around the nucleus in waves.

The number of protons in an atom determines what element it is. Every atom in the universe with one proton is a hydrogen atom, while every atom with six protons is carbon. Hydrogen is by far the most common element in the universe. The number of neutrons any atom possesses determines what isotope it is. Hydrogen has three isotopes. Its most common isotope has no neutrons; the isotope with one neutron is called deuterium; and the isotope with two neutrons is called tritium. All three isotopes are still hydrogen.

Next, we must investigate the electrical properties of the atoms. Protons and electrons have a property called electric charge. Their charges are "opposite," meaning they attract each other, while like charges, such as a pair of protons, repel each other. Protons are arbitrarily said to have positive charges, while electrons have negative charges. Neutrons are, no surprise, electrically neutral. Their role in the universe is to help bind protons in nuclei. Without them, the repulsive force between protons would prevent them from remaining bound together, thereby preventing all elements other than hydrogen from ever forming. Interestingly, without protons, neutrons wouldn't exist. Remove a neutron from a nucleus and about fifteen minutes later, that neutron will spontaneously split into a proton and an electron.

The particles interact with each other via the four fundamental forces known in nature: the electromagnetic force; the strong force; the weak force; and gravitation. The weak and strong forces just act within the atomic nuclei. The strong force binds protons and neutrons together, while the weak force is responsible for the instability of some nuclei. The weak force causes some atomic nuclei to spontaneously break apart, a process we call radioactivity. The weak and strong forces will not play explicit roles in this book, so I will say no more about them. However, electromagnetism and gravitation are key players, so here are brief refreshers on them.

ELECTROMAGNETIC RADIATION

Visible light is one form of electromagnetic radiation. This radiation is composed of particles called photons. However, these are not solid clumps, like billiard balls. Rather, they are composed of packets of waves (figure 1.7); hence, photons are commonly called wave packets by scientists. Three properties of photons are relevant:

1. They all travel at the same speed, called the speed of light (denoted c, it is about 3.00×10^5 km per second (1.86×10^5 [186,000] mi per sec).
2. Each photon has a well-defined wavelength (distance from one peak to the next, as shown in figure 1.7).
3. Photons with different wavelengths carry different amounts of energy. The shorter the wavelength, the more energy the photon contains.

Different colors are simply our brains' interpretations of visible light photons with different wavelengths. Going from longest to shortest wavelength, the colors used in astronomy are red, orange, yellow, green, blue, and violet (we don't use indigo, which Isaac Newton included in defining the colors that he proposed make up white light, so that there would be seven—that number was supposed to have important mythical meaning). This raises the question: if red are the longest wavelength visible photons and violet the shortest, could there be photons with wavelengths longer than red and shorter than violet that we cannot see? The answer is yes. As first discovered in 1800 by British astronomer William Herschel (1738–1822), there are photons with wavelengths longer than red, which we now call infrared radiation. Our bodies evolved the ability to detect them as heat.

Even longer-wavelength photons exist. Predicted by Scottish physicist James Clerk Maxwell (1831–1879) and first generated in the laboratory by German physicist Heinrich Hertz (1857–1894), they are called radio waves. By definition, radio waves are the longest electromagnetic waves, with wavelengths longer than the longest wavelength infrared (photons with wavelengths on the boundary between infrared radiation and radio waves are sometimes called microwaves).

There are also shorter photon wavelengths than violet. The first range of shorter-wavelength photons are called ultraviolet. German scientist Johann Wilhelm Ritter (1776–1810) discovered ultraviolet radiation in 1801, a year after Herschel published his discovery of infrared photons. Our

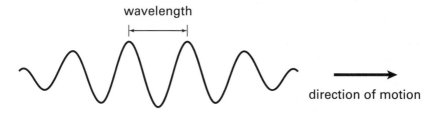

FIGURE 1.7

Schematic of a photon, with its waves of equal wavelength (distance from peak to peak as shown by the double-headed arrow).

Neil F. Comins

bodies also have built-in detectors for the longest-wavelength ultraviolet photons, called UV-A. These photons cause us to produce Vitamin D and to generate melanin, which darkens our skin to protect us from this radiation and from the shorter-wavelength ultraviolet photons, denoted UV-B and UV-C, which damage unprotected living cells. Harm from ultraviolet radiation includes cataracts, snow blindness, skin cancer, numerous other skin diseases, and DNA damage.

The next shorter photons, X-rays, were discovered in the 1880s, when they fogged up early photographic plates. They were first studied by German physicist Wilhelm Conrad Röntgen (1845–1923) in 1895. Finally, the shortest-wavelength, most energetic photons of all are called gamma rays, discovered by French scientist Paul Villard (1860–1934) in 1900.

Our bodies have no constructive responses to UV-C, X-rays, or gamma rays. We do detect UV-C and X-rays by the damage they cause to living things at sufficiently high doses. At even lower doses, gamma rays will do the same, and at sufficient levels, they quickly cause death. Figure 1.8 summarizes the wavelength and energy information about all photons.

GRAVITATION

The other fundamental force in Nature that directly affects space travel is gravitation, commonly called gravity. It is the only universal force of attraction—everything gravitationally attracts everything else. Gravitational attraction depends on just three things: the masses of the two objects involved, their shapes, and the distance between them.

Gravity is, of course, what holds things down on the Earth. As noted earlier, even the atmosphere, which is gas moving too fast to be solid or liquid on the planet's surface, is mostly prevented from floating into space by Earth's gravity. When the wind blows leaves into the air, the Earth's gravity also prevents those leaves from flying off into space, and it eventually brings them back down to the Earth's surface when the wind abates. As noted earlier, while being pulled Earthward by gravity, some things, like the *International Space Station*, don't fall down because they are moving parallel to the Earth fast enough to "miss it" as they fall. Such quickly moving objects are said to be in orbit.

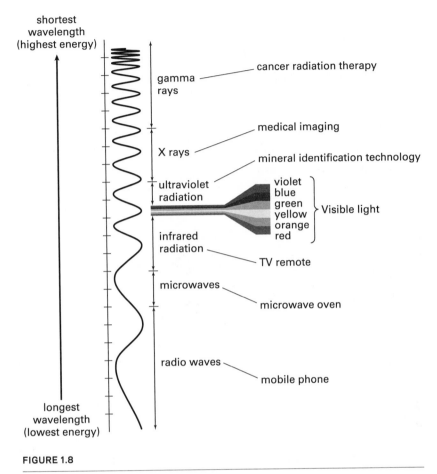

shortest
wavelength
(highest energy)

gamma
rays

cancer radiation therapy

X rays

medical imaging

mineral identification technology

ultraviolet
radiation

violet
blue
green
yellow
orange
red
} Visible light

infrared
radiation

TV remote

microwaves

microwave oven

radio waves

mobile phone

longest
wavelength
(lowest energy)

FIGURE 1.8

Summary of the types of electromagnetic radiation and their properties.

W. H. Freeman & Co.

Consider now the path of a rocket being launched from Earth. The rocket moves upward at an angle, rather than going straight up. The component of its motion parallel with the Earth's surface is necessary so that it doesn't just fall straight back down to Earth when it runs out of fuel. If it is moving fast enough, it will be inserted into orbit. If it is not moving fast enough to be inserted into orbit, then when its fuel runs out it will continue coasting upward until Earth's gravity stops its vertical climb. It then starts falling Earthward. Ignoring air friction, during all this unpowered time, called ballistic flight, the rocket follows the same parabolic path as a kicked soccer ball. The same path can be created in the atmosphere by a jet plane, as shown in figure 1.9. The plane is powered on the left side, but the engine are throttled back in the region labeled "microgravity," where its path is parabolic. The reason for this label will become clear shortly.

Air friction actually has a nontrivial effect, changing ballistic trajectories in predictable ways. Speaking of air friction, it is worth noting that rockets do not need air to operate. In other words, the exhaust from a rocket, which propels it, does not push against air to enable it to move.

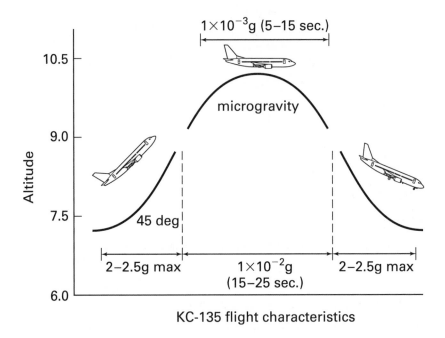

FIGURE 1.9

Parabolic flight. The central region of this figure, labeled "Microgravity," is the region where a rocket or plane that is not being propelled by an engine follows a parabolic path. The reference to microgravity and the numbers refer to training you will have in an aircraft similar to the one depicted here.

NASA

Indeed, rockets work more efficiently in the vacuum of space, where they don't have to push the air aside in order to move, than they do while climbing out of our atmosphere. The extremely low density of air in space means we don't need to worry about its effects hereafter.

SCIENCE AND SCIENCE FICTION

Sound is the compression and rarefication of air. Compression is the packing together of gas atoms and molecules, and rarefication is just the opposite—thinning of the air compared to normal. Another way of saying this is that sound is changes of air pressure. Therefore, the presence of air is essential for sound to travel. In space, the air is typically 10^{25} times less dense than the air we breathe—in other words, there are only a few atoms or molecules per cubic meter (or yard) in space. Therefore, when a spacecraft, like *Nostromo* in the film *Alien,* is rocketing through space or blowing up, there is virtually no air to carry these sounds. As a result, space is a very, very quiet place, and the sounds you hear in space in science fiction movies don't really occur.

2

BRIEF DESCRIPTIONS OF
JOURNEYS THROUGH SPACE

—

There are seven groups of space travel options available now or in the near future. As we have seen, the shortest trips will take you into space, but do not achieve orbit around the Earth—they are strictly up and back. Next are those that go into orbit around the Earth but proceed no higher. The third are trips to our Moon. Fourth are trips that leave Earth's orbit for bodies in Near Earth Orbit and will eventually return to Earth. Fifth are trips to Earth's Trojan asteroid. Sixth are trips to the moons of Mars. Finally, trips to Mars may either be one-way or round trips, although the latter are likely to be especially challenging both logistically and financially. In this chapter we will consider the process of traveling to and returning from all of these possible destinations.

Flights into space that don't get into orbit are called suborbital flights. Since they are the cheapest and shortest trips, I will explain them first.

SUBORBITAL TRIP

A suborbital flight is a three-stage roller-coaster ride. It begins with the liftoff stage, during which you will experience extreme weight; this is followed by a short period of weightlessness, and finally a second period of extreme weight as you return to Earth. As the rocket into which you are strapped lifts you toward space, you will rise faster and faster. This will have the effect of pushing you harder and harder into your seat, causing you to feel heavier and heavier. In other words, the rocket's upward acceleration has exactly the same effect on your body as does the Earth pulling you down. As unintuitive as it sounds, the gravitational attraction of the Earth causes us to accelerate downward, even when we aren't moving, while whatever we are standing, sitting, lying, or floating on is applying an equal upward force, so that we don't fall.

As you rocket upward, the Earth *pulls* you down into your seat (as usual), while the rocket's acceleration *pushes* you in the same direction, downward. Under the influence of these two forces, you are pushed down on the seat more than usual; hence you feel heavier than usual. The faster you accelerate upward during your ascent, the more you weigh during that time. You will find it hard to raise your arms during this part of the trip, because they will be pushed down against the armrests with three or four times the force that they normally experience on Earth.

The amount of acceleration from the Earth's gravity we normally experience in everyday life is colloquially called 1g. This expression has its origins in the universal practice of scientists who calculate the effect of gravity to assign the number that is its everyday acceleration near the Earth's surface, 9.8 m/sec/sec (32 ft/sec/sec), the letter g.[1] The human body can normally withstand extended periods at about 3 to 4 gs, which is what you will experience as you are rocketing into space.

As a quick aside, your weight is a measure of the force with which you are being pulled (by Earth or another world) or pushed (by the rocket) when you are at rest on the surface of any world or sitting in a rocket, respectively. Force and acceleration are directly related to each other: double the acceleration anything undergoes, and you double the force acting on it. We write this mathematically as $F = ma$, where F is the force something feels, m is its mass, and a is the acceleration it is undergoing as a result of that force. As noted earlier, mass is just a measure of the total number of particles something contains. So, standing on a scale on the Earth, you can see that the Earth exerts a force downward on you that is equal to your weight.

When nothing is preventing you from falling, you are weightless and are said to be in free fall. When you jump off a diving board you are in free fall, even while you are initially moving upward. Your path from the diving board until you hit the water will be a parabola, as described in chapter 1. Likewise, after your rocket stops firing (colloquially called flameout) on your suborbital trip, the spacecraft you are in will no longer be pushing you. Hence you will become weightless.

There are in principle three options for vehicles to take you on suborbital spaceflights. The first is planes that take off horizontally using rockets to push them up into space. After running out of fuel, they follow a parabolic orbit and then coast back to Earth.

The second option is to use a spacecraft perched on top of traditional rockets. That is how the old *Mercury*, *Gemini*, and *Apollo* spacecraft were carried into orbit, and also how the Russian *Soyuz* spacecraft is launched today and the American *Orion* spacecraft will be launched in the near future. For suborbital flights in this mode, a spacecraft would be carried aloft by a rocket, which would decouple from the spacecraft when the rocket fuel is used up. The spacecraft would then continue upward in a parabolic flight, either gliding back to a landing, if it has wings, or landing under parachutes, like the *Soyuz* space capsules.

The third option is to strap a rocket-propelled space plane under a larger plane, called the mothership, which takes off like a normal plane, lifts the smaller vehicle as high as possible, and then drops it. After being dropped, the space plane fires its internal rocket and heads further upward. After it flames out, the plane flies its parabolic trajectory and then coasts home.

Since the 1940s it has been clear than the most efficient way to fly suborbital space planes into space is the third option. This is because the mothership does the heavy lifting at no fuel cost to the smaller space plane. Furthermore, virtually all components of all the vehicles are reusable. Perhaps the most famous of these earlier craft was the X-15, which was dropped by B-52 motherships at altitudes of nearly 15 km (9 mi). Enormously fast, the X-15 flew as high as about 108 km (67 mi), well above the Karman line. Like the X-15, space planes dropped from motherships will not carry expendable parts (such as external fuel tanks) and will not need to be greatly refurbished before they can be reused. So, allowing that all three launch technologies are feasible, it is likely that most commercially available suborbital space trips will be done by dropping a rocket-equipped space plane from a mothership.

From the time you start your parabolic motion (see figure 1.9) until the atmosphere starts affecting your spacecraft as it descends, you will be weightless. Weightlessness is also called microgravity. Microgravity is the most fun part of a suborbital flight. You will be weightless for roughly four to six minutes. Assuming that there is enough room, you will be able to float in the cabin of your space plane during this time.

It is very likely that suborbital tourist spaceships will be winged craft that will return to Earth by landing like an airplane. On the downward leg, once the wings start biting into the atmosphere and lifting your spaceship (slowing the descent), you will gently fall to the floor of the vehicle and begin to feel your weight increase. At this point, the parabolic, microgravity part of your trip will be over and you will need to return to your seat. As the wings slow the plane more and more, the spaceship will be pushing up on you harder and harder. Combining this upward force of the spacecraft and the downward pull of the Earth, you will again experience 3–4 gs, as you did during your ascent into space. This extra upward force will decrease as the plane settles down to normal flight and descends gradually to a landing.

Total round-trip suborbital flight time is estimated to be between 30 minutes and 2 hours.

EARTH ORBIT TRIP

Space travel flights into Earth orbit and from there to an orbiting space resort will give travelers the opportunity to enjoy spectacular views of the Earth and indulge in exotic romantic encounters, among other things.

The ascent to orbit will probably be similar to the current trajectory taken by the *Soyuz* spacecraft on their way to the *International Space Station*. The burn (rocket on) stage to put you into orbit will last only about 8.5 minutes. You will initially be in Earth orbit at an altitude of about 240 km (150 mi). Space stations, however, are typically put into orbit much higher than this because Earth's atmosphere can be heated enough to rise to that 240 km height. The resulting air friction up there is sometimes so great that it can cause orbiting craft to lose energy, spiral Earthward, and burn up in the atmosphere. Keeping a space station in orbit is cheaper if you don't have to boost it back up too frequently. The *International Space Station* orbits 400 km (250 mi) above the Earth's surface. This is part of the realm of space called low Earth orbit, which begins at the Karman line and extends upward to about 1,900 km (1,200 mi) above Earth's surface.

Getting to your Earth-orbiting destination is a two-step process. After the first step of getting into orbit, well below the orbit of your destination, your spacecraft will fire small rockets that will cause it to spiral farther away from the Earth. Assuming, plausibly, that your Earth-orbiting destination is located at about the same altitude as the *International Space Station*, it will ideally take about five hours for your ascent spacecraft to climb the additional distance and dock with your destination station. If there are any technical glitches, such as your initial orbit being slightly different than planned or equipment that needs repair, then it can take up to two days to reach your destination.

The orbital path of your destination space station is important in determining what you will see from orbit. At an altitude of 400 km, objects orbit the Earth once every 90 minutes. Therefore, if you are orbiting directly over the Earth's equator, then every 90 minutes you will be over the same place you were 90 minutes before. There are lots of fascinating things to see in the latitudes around the equator, but on such an orbit you would be unable to see the regions located at high northern and high southern latitudes. For example, you can't see London from equatorial orbit at an altitude of 400 km.

To get the most bang for your buck, you will want to see as much of our extraordinary planet from space as possible. Your destination station will therefore not orbit over Earth's equator. Rather, like the *International Space Station*, your destination is likely to cross Earth's equator at an angle of about 50° (figure 2.1). As a result, you will be flying over places as far north as 50° north latitude (including London) and as far south as 50° south latitude (including Melbourne, Australia).

Another advantage of being at this low altitude, besides the relative ease of getting there and the rapid orbits around Earth, is that you will be relatively safe from the high-energy solar wind particles trapped above that altitude by the Earth's magnetic fields. Earth is surrounded by these magnetic fields (think the attractive and repulsive forces from bar and horseshoe magnets that you may have played with, and the little magnets that hold notes on your refrigerator). Many high-energy (high-speed) protons and electrons are trapped, primarily in two regions. These particle-

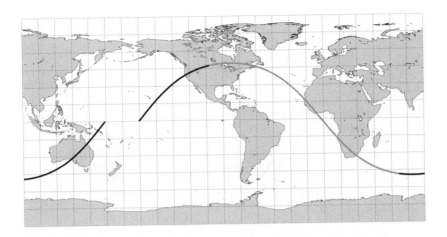

FIGURE 2.1

The curve shows nearly one orbit of the *International Space Station,* moving from left to right. This path starts at the equator southwest of the United States and ends at the same latitude northeast of Australia. Note that the *Station* crosses the equator at different places after each orbit. As a result of this shift, each orbit is different; you will eventually see virtually all of Earth's surface. The light region of the path is daylight.

rich regions are called the Van Allen radiation belts after James Van Allen (1914–2006), who instrumented early rockets with Geiger counters that detected the radiation in 1958 (figure 2.2).

The trapped particles in the Van Allen belts are moving so fast that they can penetrate space suits, shielding, and outer layers of spacecraft and satellites, making extended travel in those areas extremely harmful to life and equipment. The inner belt normally extends outward from about 965 km (600 miles) above the Earth's surface. However, there is a region of this belt, called the South Atlantic Anomaly, where it descends to about 190 km (118 mi) above the surface (see figure 2.2). The *International Space Station* passes through this region about five times daily, spending up to 23 minutes in the inner Van Allen belt. Astronauts report seeing "shooting stars" in this region, due to high-energy particles there passing through their eyes and brains.

After your space station experiences, which might include a space walk, you will either return to Earth in a capsule that would parachute onto land, as the *Soyuz* capsules do today, or, given the success of the

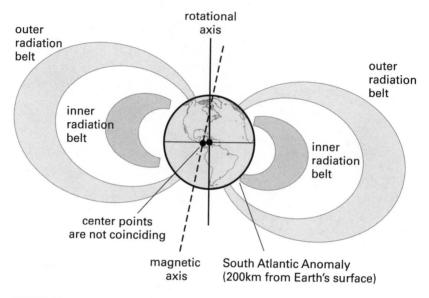

FIGURE 2.2

Cutaway of the Earth and the two major Van Allen belts, which are magnetic fields containing large quantities of high-energy charged particles streaming Earthward from the Sun.

W. H. Freeman & Co.

space shuttle, return in a winged spacecraft bringing you back to a normal aircraftlike landing.

Total round trip could run from a week to months.

TRIP TO THE MOON

The Moon was an ideal goal for early human spaceflight. In the first place, mid-twentieth-century science and engineering were sufficient to send astronauts there. Furthermore, its low mass compared to the Earth allowed that same engineering to craft rockets that could lift the astronauts off the Moon and back into orbit around it, so they could be picked up and brought home. (Similarly, the former Soviet Union sent unmanned space-craft there, three of which brought samples of the Moon's surface back to Earth.) If the Moon had the mass of Mars, say, then the gravitational attraction on its surface would have been too great for early technology to bring astronauts back home. We will explore this latter issue further when we discuss travel to Mars today.

Getting to the Moon and back has been a messy undertaking. When each set of *Apollo* astronauts went to the Moon, NASA intentionally wasted a considerable amount of the hardware that went up with them. First they rocketed into Earth orbit. The lower rocket stages were jettisoned and burned up as they reentered Earth's atmosphere. From orbit, a booster rocket sent the astronauts in an initially four-piece spacecraft rapidly through the Van Allen belts into a "translunar trajectory." Then the booster that got them out of Earth orbit was also jettisoned. What remained were the command module, the service module, and the lander/ascent system. (The lander/ascent system carried two astronauts to the Moon's surface and then back into lunar orbit.) As the ship approached the Moon, a booster on the service module ignited, slowing its motion and locking it into lunar orbit. Finally, the lander/ascent system decoupled from the command and support modules and descended to the surface, slowed by a rocket in the lander.

After exploring a small region of the Moon, the two astronauts down there entered the ascent module, which blasted off and carried them back to the orbiting command/service module, leaving most of the lander on the Moon's surface. The orbiter and ascent module reunited and the two astronauts and their cargo reentered the command module. At this point, the ascent module was jettisoned and the service module rocket was fired again, taking them out of orbit and back toward Earth. On the way, the service module was also jettisoned (it too burned up in the Earth's atmosphere), and the three astronauts rode the command module down to a parachute landing in the ocean, where they were picked up by a helicopter.

Clearly, the *Apollo* scenario was a very inefficient use of hardware, since the command module represented only a few percent of the total mass of the system that left Earth orbit in the first place. While that was not a problem for the seven *Apollo* missions, it would be economically unacceptable for future space travel. That is why there will likely be reusable spacecraft for all stages of the journey to the Moon and back when you go.

Getting to the Moon takes several steps. First you will be taken to low Earth orbit, as described above. You may stay on an Earth-orbiting station for a few days, acclimating to microgravity. From Earth orbit you will

embark on a shuttle to lunar orbit. Critical to leaving Earth orbit is getting through the Van Allen belts as quickly as possible, to minimize your
exposure to the radiation there. Once you are beyond this region, you will
coast to the Moon. This trip from Earth orbit to lunar orbit will take about
three days, during which time you will be weightless except for the few
minutes when you rocket out of Earth orbit and again for a few minutes
while you decelerate (slow down) into orbit around the Moon. The docking process will bring you either to a space station that has lunar landers
waiting for you or directly to a lunar lander orbiting the Moon. In either
case, the lander will bring you down to the Moon's surface.[2] Returning to
Earth is likely to be the reverse of this process.

**Total transit time from Earth to the Moon or back will probably be
between three and seven days. Total time on the Moon could vary from
a few days to months.**

TRIPS TO ASTEROIDS AND COMETS

Depending on when you go beyond the Earth-Moon system, you will have
a choice of several Near Earth Objects to visit. However, trips to different Near Earth Objects will come with seriously different costs, both in
money and in time spent on the journey. There are two reasons for this.
First, all the Near Earth Objects except the Trojan asteroid take different
lengths of time to orbit the Sun than does the Earth. This means that they
are only rarely in locations relative to the Earth that are easy to get to in a
reasonable amount of time, meaning weeks or a few months.

Second, many of these objects do not orbit the Sun in precisely in the
same plane as the Earth, called the ecliptic. We normally send spacecraft
to other planets or small solar system bodies orbiting in or very near this
plane. It takes much less energy to put a spacecraft into an orbit on the
ecliptic than to put one into an orbit that significantly leaves that plane.

The Hohmann transfer orbit is an efficient trajectory to travel destinations beyond the Earth-Moon system. These flights involve just two fir-
ings of your ship's rocket, as shown in figure 2.3.

Leaving Earth orbit is much easier than leaving the surface of the
Earth. A much less powerful rocket than the ones that lifted you into

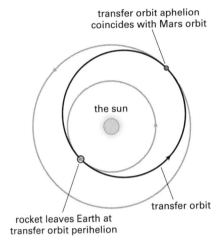

transfer orbit aphelion
coincides with Mars orbit

the sun

transfer orbit

rocket leaves Earth at
transfer orbit perihelion

FIGURE 2.3

The Hohmann Transfer orbit from Earth to Mars. This same trajectory applies to going from Earth to any body in a near-circular orbit in the plane of the ecliptic.

NASA

orbit is fired to give your spaceship energy so that it spirals outward from Earth orbit. The rocket is then fired again as you approach your destination to insert your spaceship into the same orbit as your target world. Without requiring further course corrections, this only works for bodies in nearly circular orbits in the plane of the ecliptic. To get to objects that are orbiting in planes tilted slightly compared to the ecliptic—as are many of the target worlds for upcoming human space travel—course correction burns need to be added to the Hohmann transfer orbit.

Another near-Earth trip is to Earth's one known Trojan asteroid, presently labeled 2010 TK_7, which leads the Earth in its orbit around the Sun by 60°. Travel to it will require a different but simple trajectory. A small change in the spacecraft's speed and direction compared to that of the Earth will carry the craft ahead of our planet and toward the asteroid. This asteroid, which is about 300 meters (1,000 feet) across on its longest diameter, will eventually be given a Trojan-related name. Here is space to fill in the name it is given: _____ .

Obviously, having 2010 TK_7 always in the same place relative to the Earth makes getting to its vicinity convenient. In principle, you could go whenever you want, compared to traveling to comets or Apollo, Aten, Amor, or Atira asteroids, which are each near Earth at only certain times, requiring your departure to be cast in concrete months or years in advance.

However, 2010 TK$_7$ does not orbit in the plane of the ecliptic. Instead, like a porpoise following a boat, this asteroid moves above and then below the ecliptic as it orbits the Sun. While getting a spacecraft near it is relatively inexpensive, its up-and-down motion adds to the fuel expense required to bring a craft into its orbit so it can land. The obvious thing, to wait until the Trojan asteroid crosses the ecliptic where your spacecraft is normally traveling, is tricky because the asteroid will be rising above or diving below the ecliptic at that time. Your ship would have to be equipped with a booster rocket to match the asteroid's speed perpendicular to the ecliptic so you could visit it, and then another booster rocket to return you to an orbit back on the ecliptic. As noted, the issue is money, rather than any technological problem. It is entirely likely that there are other Trojans orbiting with Earth, and some of these may well orbit very close to the plane of the ecliptic. That being the case, they would be relatively economical targets for space travel this century.

■ ■ ■

Knowing the orbits of Apollo, Aten, Amor, and Atira asteroids, and short-period comets, astronomers can predict when they will be passing close enough to Earth to make them logistically and financially plausible targets for space travel. In the early days of visiting these bodies, I envision some tourists going along for the ride with either research astronauts going to study them or workers going to mine them. This scenario is reminiscent of travelers on tramp steamers; the ships' main purpose was to deliver cargo, but they would take people willing to pay for the voyage.

Your travel to an asteroid or comet can take one of two routes. Both begin with a trip into Earth orbit. Then you would be boosted out of orbit on a trajectory that would cross that of your destination body. The other option being explored is to go from Earth orbit into orbit around the Moon, and from there out to the destination. Aerospace scientists are considering this latter path because it will take less energy, which translates into less money, to leave lunar orbit and go out into interplanetary space than to leave Earth orbit directly. Likewise, returning from deep space can sometimes be done more efficiently by going first into lunar orbit and from there to the Earth.

In summary, astronomers are assembling a list of asteroids and comets that are plausible exploration, mining, and tourist destinations. The comets, especially, will be moving targets for trip planning because relatively few of them orbit in the plane of the ecliptic, and the orbits of those that do are typically changed each time around the Sun by gravitational tugs of the planets and dwarf planets that they pass.

The length of your trip to a passing asteroid or comet will depend on the orbit of your destination relative to Earth's orbit around the Sun. You can plan on months to a year or more for these trips.

TRIPS TO MARS OR ITS MOONS

Trips to Mars and/or its moons are the ultimate space voyages for humans in the near future. As we have seen, getting to the vicinity of Mars begins with a trip from Earth to low Earth orbit. It is most economical to leave from there directly for the Red Planet. As shown clearly in the movies *2001: A Space Odyssey* and *The Martian*, for example, it is possible to create the equivalent of gravity in a spacecraft for a voyage to Mars. This is done with an oversized centrifuge built into the spaceship, which slowly spins around, forcing the occupants to its outer edge. Spinning at the appropriate speed, you will be pushed outward with 1 g, so that you can stand on the outer edge and feel your normal weight. However, this is *very* expensive, so it is likely that early trips to Mars will be done with travelers floating in microgravity.

Your ship could take a Hohmann transfer orbit to Mars, or it could take an even more cost-effective ballistic capture trajectory, which lets the gravitational attraction from Mars on your ship help pull it into orbit. This has the advantage that rockets do not have to be used to slow down the spaceship as much as with the Hohmann transfer orbit, but the disadvantage of taking longer.

Until the giant centrifuges become economically feasible, you will be weightless once you leave Earth orbit until you start slowing down near Mars. As discussed earlier, the different orbits of Earth and Mars around the Sun strongly affect the length of time it takes to get there or back to

Earth. The relatively light spacecraft that we send to Mars now take between about four months (*Mariner 7*) and nearly a year (*Viking 2*) to make the journey. You can expect transit times of six months to a year to Mars and about the same length of time coming home.

Let's first consider trips to the moons, of which Mars has two. Unlike our Moon, they are both very small compared to their planet and irregularly shaped (figure 2.4). Whereas our Moon is about 1.2×10^{-2} times the mass of Earth, Phobos is 1.7×10^{-7} times the mass of Mars; Deimos is even less massive. Indeed, they are likely to have been asteroids that were captured by the planet. Phobos is everywhere less than 27 km (17 mi) across and only 6,000 km (3,700 mi) above the surface of Mars, while Deimos is about 16 km (10 mi) across and 20,000 km (12,500 mi) above the planet's surface. They are orbiting so close to Mars, compared to our Moon's orbit around the Earth, that the closer one, Phobos, orbits every $7\frac{2}{3}$ hours, while the farther one, Deimos, goes around about once every $30\frac{1}{3}$ hours. For comparison, our Moon orbits once about every 27 days. We have not yet sent any spacecraft onto either Phobos or Deimos. Nevertheless, images taken by Mars orbiters show them to have craters and powdery regoliths. It would take a relatively trivial amount of energy to land on and take off from either of these moons.

At this point, you may understandably be thinking: *Why travel all the way to Mars only to visit its moons?* I can't give you an exciting sales pitch about the moons, such as, "They are mostly made of gold and you can bring some of it back with you." We just don't know yet, but that doesn't mean that when we explore them scientifically we won't find something appealing for other travelers. The reason that travel to just the moons may be desirable now is more prosaic—we can actually reach them and reliably return from them. In comparison, journeys onto and especially back off of Mars are by far the most technologically challenging and hence the most expensive of all planned spaceflights in the foreseeable future.

Consider the issues related to landing on and leaving the Red Planet. Mars is roughly half the diameter of the Earth, twice the diameter of our Moon. The gravitational attraction at the surface of Mars (how much you would weigh there) is about 0.4 times as much as on Earth or about 2.3 times as much as on our Moon. The "So what?" about all these numbers is that it is much more difficult to land on Mars, which is pulling your

FIGURE 2.4

Mars's two moons, Phobos (left) and Deimos. Crater Stickney covers most of the right side of Phobos in this photograph.

NASA/JPL-Caltech/University of Arizona

lander down harder, than it is to land on our Moon or any other space destination in the near future. For the same reason, it is much harder to get back into space from Mars than from the Moon.

Let's consider landings first. The harder a body pulls your lander down, the more powerful your rocket and other landing equipment have to be to slow you enough for a gentle touchdown. The challenge of landing on Mars is further complicated by the fact that unlike our Moon, Mars has an atmosphere. The air pressure on Mars is about 0.007 times as great as the air at the surface of the Earth. Put another way, the air on Mars is 60 times less dense than the air we breathe. (Even if it had the right chemical composition to be breathable, which it doesn't, that low density would not provide enough oxygen for you to survive.)

The low air density and air pressure on Mars will affect the process of getting onto the planet's surface. Parachutes are an integral part of virtually every landing on Mars. Since the air gets denser the lower the surface is, landings are typically done in valleys, where the denser air can slow parachutes most effectively. However, even where it is densest, there isn't enough air surrounding Mars to enable landers to land gently any-where just using parachute technology. Soft landings require additional equipment, such as rockets that fire after the parachutes have initially slowed the descending spacecraft.

Because of the challenges associated with landing on Mars, it is un-likely in the foreseeable future that landers there will be qualified to take off again. Instead, as science fiction films such as *The Martian* correctly portray, it is likely that a separate set of return rockets will be landed with great care before the people who will use them to leave the surface

SCIENCE AND SCIENCE FICTION

Visualize yourself picking up a balloon that has not been inflated. The rubber is soft and flexible. Now start blowing it up. It takes some effort to exhale enough air to force the rubber to expand. Eventually, you will have a relatively rigid rubber object in your hand with enough air in it, meaning with sufficiently high air pressure, that if you let go of it, the air will rush out as the balloon flies around the room.

Implicit in this scenario is that you were breathing air at normal air pressure of 100 kPa (kPa is shorthand for 1,000 Pascals, commonly called a kilo Pascal), or 14.7 pounds/square inch. This pressure of air composed of the same molecules that we breathe is roughly what people in spacecraft will be breathing. Indeed, that is the air pressure in the *International Space Station* and is likely to be the standard pressure for most spacecraft and habitats off Earth. However, in the movie *Mission to Mars*, the rescuers arrived to find character Luke Graham living in a greenhouse (useful for growing plants that make oxygen from carbon dioxide, as well as provid-ing food). The walls of the greenhouse were fabric fluttering in the breeze. The problem is, as discussed earlier, that the air pressure on the surface of Mars is about 160 times lower than it is here or than it is inside the greenhouse in the movie. Therefore, the greenhouse walls should have been expanding out like a balloon, rather than fluttering in the breeze. They would burst if not made of a very strong fabric.

even arrive. The reason is to ensure that the return rockets are in working condition so that the people who follow and want to go home can do so.

Problems landing the return rockets start in the atmosphere, where air friction heats them up. This can damage both mechanical and electrical equipment. The buffeting of the air can also cause the rockets to flex, which would tear electrical cables and crack metal. If the touchdown of a return rocket is anything but extremely gentle, the impact on the surface will likely cause serious damage too. It takes just one malfunction to cause a rocket to fail on launch or go dramatically off course on the way up, killing the passengers on board.

The good news about getting off Mars is that it is an engineering issue entirely; there are no purely scientific hurdles to overcome. Putting enough money into the project will lead to development of an ascent rocket that can be reliably landed on Mars and then be refueled and used to lift people off the Red Planet. This challenge is the reason some people in the space travel business are considering one-way trips to Mars in the relatively near future and explains why it will be so much cheaper just to visit Phobos and Deimos. Having said all that, in chapter 10 I will still explore the opportunities for round-trip tourists to Mars, as well as for people who emigrate there.

3

PREPARING FOR YOUR TRIP

———

Planning for a long-distance trip on Earth today is relatively straight-forward. For example, my wife and I have never been to Ireland. A viable scenario for planning such a trip might look like this: after conversations with friends who have been there and a few hours surfing the web, we have convinced ourselves that we want to visit the Emerald Isle. We see on the computer that there are any number of travel, tour, lodging, and dining options. With a few clicks of the mouse, we've got our reservations. I check on visa requirements and our vaccinations. Then we look at weather forecasts for the period of the trip, which will determine what clothes we will pack. I make sure that our passports are up to date; that our cell phones will work there; and that we will have all the medi-cines we will need to take during the trip. We may read up on historic places in Ireland and buy some actual guidebooks or stream some on our smart phones or tablets. Ideally all this takes a few hours or days, and we are basically ready to go.

Your journey into space will require a bit more planning and research. The first step, once you have chosen a destination, is to find a carrier to take you there. Like booking travel on Earth, this can be done online. That is the easy part. Now the hard work begins. Unlike people booking a flight to Ireland, everyone planning to travel in space must be screened for a variety of medical, dental, and psychological problems that would prevent them from going.

Your medical history will be key. Research about human adaptation to space is limited by the exceedingly small number of people who have traveled up there. Physicians really don't yet know the effects of all preexisting illnesses on space travel or the effects of space travel on all illnesses (either preexisting or that come on while in space). Furthermore, not all the space travelers to date have stayed for the same length of time, have flown in even remotely similar spacecraft, or have even reported all the illnesses that they experienced. It is therefore very difficult to ascertain which illnesses are actually caused by space travel.

MEDICAL AND PSYCHOLOGICAL TESTING FOR SPACE TRAVEL

Nevertheless, certain requirements for space travelers are being codified. The first standards for people wanting to visit the *International Space Station* were published in 2007 by the Center of Excellence for Commercial Space Transportation and updated in 2012. Although these were funded by the U.S. Federal Aviation Administration, they are presently unofficial standards. Further updates will likely follow. This process of setting standards draws on experience from three settings. First, normal social, medical, and psychological experiences in everyday life provide some basic guidelines for who can and who should not go into space. For example, people with certain social disorders that prevent them from interacting well with others on Earth are not permitted to go into space, since they are likely to create potentially dangerous situations up there. Similarly, people with a variety of medical problems that cannot be effectively remedied are barred from travel on the *International Space Station*. Such conditions include certain heart conditions; physical deformities that prevent them from using essential flight equipment such as space suits and helmets; uncorrectable hearing problems; vision problems; blood disorders; respiratory and digestive problems; metabolic diseases such as diabetes and gout; thyroid problems; and kidney problems.

For female travelers, standards are even more restrictive. Pregnancy and the safety of human egg cells in space are serious concerns. Although no human fetus has developed in space, it is believed that fetal development in microgravity and high levels of radiation would lead to serious problems for both the fetus and its mother. Thus, unless and until we learn otherwise, pregnant women cannot travel in space.

The psychological standards we have are based in large measure on Earth equivalents of spacelike conditions and how people fare in extreme environments. To this end, researchers study people in high-stress, high-performance, often isolated environments such as occur in submarines, cave exploration, Antarctica, and military aviation.

Even when people are medically and psychologically fit for spaceflight, the experience may affect them after they return to Earth. The bodies of medical, social, and psychological knowledge that are being gathered from all the people who have been in space are growing and becoming statistically significant. Researchers are now identifying a number of problems that commonly develop as a result of spaceflight. This information is helping to guide decisions about present and future astronauts and cosmonauts. For example, more is known about the percentage of astronauts who have suffered glaucoma after returning to Earth compared to the percentage of otherwise similar people who get glaucoma without ever going into space. Such statistical analyses are challenging, since there are many characteristics, such as age, gender, medical history, and typical exposure rates to sunlight, that must be taken into account before meaningful conclusions about the effects of space travel can be reliably asserted. Indeed, this field of space medicine, only a few decades old, is very much a work in progress.

To give you a perspective on the medical tests you will undergo before being allowed up into space, you can read online the 2012 *Flight Crew Medical Standards and Spaceflight Participant Medical Acceptance Guidelines for Commercial Space Flight*. Some of the results of this screening may prevent you from going into space; others will just require medical care before you will be allowed to go.

LEGAL AND INSURANCE ISSUES
RELATED TO SPACE FLIGHT

It is important to be aware of legal issues surrounding spaceflight and the insurance that will be available for you. I buy travel and cancellation insurance whenever I am going to fly on an airplane. This covers missed connections, life (accidental death and dismemberment) insurance, medical care during travel that isn't covered by my normal medical insurance, theft, damage to luggage, lost luggage, and trip cancellation, among other things. It seems likely that you will want similar insurance for your trip to space.

Insurance already exists both for space travelers and for the companies that are involved in space travel. However, the issues surrounding insurance are sufficiently complex that it is worthwhile understanding some of them. In the big picture, insurance rates are determined primarily by the probability of a certain event requiring payout, such as a trip being canceled, and the amount that has to be paid by the insurer when that event does occur. Actuaries use a variety of mathematical tools, based on the relevant data for the events they are insuring, to determine how much we pay for each insurance policy.

The biggest problem for determining the rate you will pay for space travel insurance is that there have been few people in space and very few accidents or incidents in which insurance claims have been made. Furthermore, there needs to be a supply of money available to pay claims. Normally, insurance companies collect enough money from customers to pay all their claims (and still make a profit). Companies in the new field of spaceflight insurance haven't collected nearly enough money to do that yet. Either they have to rely on money from other sources to get started or they have to hope that the likelihood of events that will force them to pay out is sufficiently low that they can earn enough before they have to begin paying customers.

A few other insurance issues are significant. Suppose a rocket carrying a transport vehicle with passengers explodes on the launch pad. Several parties will require insurance reimbursement. First, the spaceflight company will have insured their hardware (rocket and launch system). Also, if they are smart, they will insure themselves against the loss of revenue that they will endure as their launch system is being rebuilt and as they and the relevant government agency, such as NASA, figure out what went wrong. During this time, their flights are likely to be grounded. Second, the owner of the payload, which may or may not be the space flight company, will have insured their vehicle. Third, crew and travelers or their survivors will have to be reimbursed. And finally, parties who are not involved in the spaceflight program but who suffer damage or injury must be compensated. These are called third parties in the trade.

Third-party liability is something spaceflight companies must have. For example, your spacecraft could take off, but a piece of it fall off and slam into someone's swimming pool, destroying it. The person who owns the pool has nothing to do with space travel and is therefore a third party in this business, with the spaceflight company and you, as passenger, being the first and second parties. The swimming pool owner has every right to expect to be compensated by the spaceflight company. Needless to say, third-party events involving death are far more serious and expensive.

Accidents have happened. On June 28, 2015, a Falcon 9 rocket owned by Space Exploration Technologies Corporation (SpaceX) carrying supplies and experiments to the *International Space Station* exploded minutes after launch. While both the rocket and its payload were destroyed, they were covered by insurance. This rocket lifted off from Cape Canaveral, Florida, heading eastward. As a result, all the debris splashed into the Atlantic Ocean, so there were no third parties injured or properties damaged by the event.

To describe some of the other legal issues surrounding your trip, suppose you took a spaceflight during which your arm was crushed by a massive container of food being transferred from the space capsule in which you also traveled to the orbiting hotel where you were staying. Can you sue the spaceflight company or the hotel? The answer is no, because before departing you would have signed an informed consent form stating that you know of the potential dangers and will not hold the spaceflight company or the hotel responsible in case you sustain an injury or death. However, depending on how much insurance you bought, your insurance will cover your expenses if you are injured or provide money for your estate if you die.

Legal issues pertaining to space travel are extremely complex, and we still have a long way to go before they are codified. Consider another example. Suppose you rode a rocket up from Spaceport America, located in southern Arizona. Something went wrong, forcing your spacecraft to make an emergency landing near Amarillo, Texas, and causing considerable damage to Interstate 40 and the surrounding farmland, as well as to the lander itself. Whose laws concerning liability does your spaceflight company have to follow? Those of New Mexico? Texas? Potter and Randall counties, in which Amarillo lies? The United States? This question is yet to be answered.

As another example, suppose a spacecraft took off from Cape Canaveral and crashed in Germany. Then whose liability laws would apply? The United Nations has the authority to decide these matters, but only if there is consensus between the countries involved. This is rarely achieved, so current laws are worded sufficiently vaguely that they can be passed and implemented without ruffling too many feathers. However, now that commercial space travel is here, these legal issues must be refined into meaningful laws. Needless to say, along with space medicine, space law is a work in progress.

4

TRAINING FOR SPACE TRAVEL

———

An adult I know, let's call him Alex, had never flown before (true story—only the name and location are changed). He wanted to go from Maine to a scientific meeting in California to which I was also going. I offered to help him though the travel process, which he gratefully accepted. First I told him how to pack his bags and what he could and could not take on board with him. At the airport in Bangor, Maine, I showed him how to check in, and then let him do it on his own. At security I explained that the folks working for the TSA have jobs to do and not to take any untoward behavior from them personally. Then I went through first, and he followed. Thereafter, at each step through the trip, I told him what to expect and how to handle it. We debriefed as necessary, and by the time we arrived at LAX, he was comfortable with the process.

Space travel is so different from travel anywhere on Earth that, except possibly for suborbital flights, I predict it will never be even remotely close to that simple. With this in mind, let's next consider what training you will have to undergo in preparation for your journey into space.

Depending on their role in the mission, a professional astronaut normally requires two to three years of training to qualify for spaceflight. Your preflight training will be much shorter. If present commercially available opportunities are any indication, it will take you a few days to train for a suborbital flight, and at least one to three months to prepare to fly in orbit around the Earth or to the Moon, depending on what kind of trip you take. Training for journeys to space debris or to Mars will take significantly longer because of the more complicated issues and longer times associated with these ventures. The differences will become apparent in the chapters to come.

The basic training in preparation for any spaceflight will begin with classroom and book learning about what you will experience, followed by hands-on training for those elements that can be effectively simulated before you depart. You will experience what it feels like to undergo extreme acceleration, microgravity, low air pressure, space suit use, getting in and out of spacecraft, using safety harnesses (fancy seat belts), as well as eating, drinking, and using the toilet in microgravity and dealing with a variety of emergencies. All this effort will prepare you for most of the experiences you are likely to have out there. Knowing what to expect will make the trip infinitely more satisfying and safer for you and the other travelers, compared to the chaos that is certain to occur if you go without this preparation.

The psychosocial aspects of your time in space are also vitally important. These involve such matters as getting along with your companions, making the most of the time and opportunities you will have, and dealing with events happening back home while you are away. These issues will be discussed with you before you leave, along with successful techniques for handling many of them. I explore these matters further in chapter 7.

Many experiences in microgravity and exposure to particles and radiation in space have no parallels on Earth, and center around how your body will physically change as you adapt to even a few days in microgravity. These matters are discussed in chapter 6. Let's now consider the hands-on experiences you will have prior to leaving that are essential for adapting to life in space. We start with pulling gs.

EXTREME ACCELERATION

You will undergo a variety of accelerations during your travel into space and back to Earth that are beyond anything you have yet experienced. Acceleration is generically a change in the speed or direction of your motion. It is common to call an increase in speed or an increase in the rate that you go around a curve an acceleration and a decrease in speed or motion around a curve a deceleration, which I will do from now on when the difference needs to be made clear. The ride into orbit and the return to Earth's surface require you to undergo extreme acceleration for about nine minutes on the way up and then extreme deceleration for about eight minutes on the way down.

What do these extreme sensations feel like? Imagine flooring the gas pedal in a hot car of your choice from rest. In 6 seconds you are traveling at 97 km per hour (60 mph). Clearly, you have accelerated throughout this time interval and as a result, you experience a force ($F = ma$) pushing you back against your seat. We compare such accelerations with what you normally feel from the Earth pulling you down, 9.8 m/sec/sec (32 ft/sec/sec). For example, this notation means that every second you are falling near the Earth, you are traveling 9.8 m/sec (32 ft/sec) faster than you were a second ago. The terminology for undergoing an acceleration or deceleration is "pulling gs." Recall from chapter 2 that one g means one times the acceleration of gravity from the Earth. Before your car starts moving, you are pulling zero gs horizontally. While accelerating, you are pulling nearly .5 gs horizontally.

Humans are relatively fragile creatures. We can only withstand a finite amount of acceleration in any direction before we are permanently injured or killed. Furthermore, the direction of the acceleration relative to your body plays an important role in determining how many gs you can safely withstand. Consider for these examples a healthy adult. If you are standing in an elevator that starts moving upward, you experience acceleration along the axis of your spine. The blood in your body is forced down toward your feet. If that acceleration is sufficiently great, about 5 gs, so much blood will leave your brain so quickly that you will pass out, not to mention that your spine will have been crushed. You will also have sustained other internal injuries.

Suppose next that you are sitting in a car and experience acceleration pressing you against the back of your seat. In this case, your blood flows from the front of your body to the back. Conversely, you could be driving at speed when something happens in front of you that forces you to jam on the brakes. You would then be pushed forward against your seat belt. This deceleration would force blood to the front of your body. In either case, the healthy adult can withstand only about 6 *gs* while sitting for extended periods (minutes) before passing out.

As a brief aside, whenever you jump or fall, you accelerate downward due to the force of gravity from the Earth. When you strike the ground, you are decelerating from whatever speed you had just before the impact to zero km/h (0 mph) upon impact. Unlike riding a rocket, this is a very short period of deceleration. The human body has limited ability to withstand such sudden, rapid changes in speed, which is why our bodies are injured if we fall from sufficient heights. Under ideal conditions of spaceflight, you won't experience sudden accelerations or decelerations, but one can envision situations when this can occur, such as a docking maneuver that goes awry.

The consequences of prolonged extreme acceleration/deceleration can be quite serious. They include passing out, breaking your spine, damaged vision, blindness, aneurisms (bulges of blood vessels), and damage to your circulatory system in general and your heart in particular, among other things.

Before going into space, your tolerance for the high acceleration you will have to endure during your ascent and descent back to Earth or onto another world will be tested. It is likely that this will be done in a giant centrifuge (figure 4.1), which is a high-tech version of a merry-go-round.

FIGURE 4.1

Centrifuge in which astronaut trainees are spun to experience high forces analogous to those felt during launch from and reentry into the Earth's atmosphere.

NASA

Typically, you will be strapped into a seat on the outer edge of the centrifuge facing inward (although different orientations are possible). Then the machine will start spinning you faster and faster. As on a merry-go-round, this acceleration will have the effect of pushing you outward. In other words, while facing inward, you will be pressed harder and harder against your seat, just as you would be in an accelerating car or rocket. Therefore, the centrifuge can be used to measure how you respond to the equivalent of a rocket launch when you would be pulling gs.

During your time in the centrifuge, you will sometimes be spinning at a rate so that you feel an equivalent of between 4 and 6 gs. The durations of these experiences will each be less than ten minutes. They are designed to give you experience with the amount of acceleration you will feel when ascending into orbit and then returning to Earth.

MICROGRAVITY

While launch, orbit changes, and landing require you to undergo increased acceleration, you will experience microgravity or weightlessness during some part of every trip into space. If you go on a suborbital flight, it will just be for a few minutes during the parabolic part of the trajectory. If you go to a space hotel orbiting Earth, you will be weightless as long as you are in orbit.[1] If you go to the Moon or another body, you will also be weightless while in transit to and from Earth. Weightlessness, exciting as it sounds, is extremely disruptive for humans. It has both short-term and long-term effects, discussed in chapters 6 and 7. Here I will just explain what you will initially experience in a weightless environment, since you will experience that in your preflight training.

Our ability to sense gravity, that is, which way is down, is built into our inner ears. There, little calcium blobs called otoliths rest on tiny hairs called cilia. Normally the otoliths are pulled down by gravity, and nerves connected to the cilia detect their responses to the otoliths. For example, when you lean forward, the otoliths move toward the front of your inner ear, bending the cilia forward, which is interpreted by your brain as "I'm leaning forward."

However, when you are weightless, your otoliths don't have any gravitational force pulling on them, so they can't supply your brain with orientation information. You then have to rely on your eyes. However, during weightlessness there is no measurable sense of which way is down that your eyes can tell your brain either. Even when signs with words, like "this way down," are posted, your brain tries to refer to your inner ears for guidance they can't give.

About three quarters of people experiencing microgravity for the first time undergo a disorientation called Space Adaptation Syndrome or space sickness. If you have ever experienced seasickness, you have a rough idea of what this is all about. The effects of the rolling and pitching of a ship at sea are roughly analogous to the neural confusion you feel in microgravity. The motion of the ship confuses your brain, which responds as you are likely to respond in space: it makes you undergo emesis (you vomit). Indeed, cruise lines put out barf bags just about everywhere on the first night of a cruise because this can hit anyone at any time early in the trip. The effects of space sickness include nausea, vomiting, extreme headaches, drowsiness, disorientation, sweating, and loss of appetite, among other things. The disorientation can cause the room to appear to be spinning around, even if you are not moving. The good news is that space sickness is usually a one-time experience lasting less than three days, which brings us to the preflight work-up you are likely to undergo in preparation for it.

If you go into space without understanding, experiencing, and preparing for weightlessness, you will very likely be disappointed with the trip. Meds can help you avoid space sickness on short trips, and you will be given them at all times when you are wearing a space suit. But when you stop taking the meds you are again susceptible to space sickness, so you will likely be encouraged to just go through it and move on. To help you understand space sickness (and possibly experience it before you go into space), you will undergo weightlessness prior to departure.

The microgravity training you will experience before going into space and during which you might feel space sickness takes place in a jet plane that flies a path similar to that of a porpoise swimming behind a boat (see figure 1.9). During the parabolic part of the flight, you are weightless. The planes used for weightlessness training, called Vomit Comets, undergo as many as 30 cycles of up and down arcs per training flight. In each cycle, you are weightless for between about 15 and 30 seconds. Then the pilot pulls the plane out of the dive, levels off, and climbs so that next cycle can begin. Even these short periods of weightlessness cause many people to vomit. You will experience about 2 gs in the leveling off and climbing phases.

Weightless experiences on Earth can also take place either underwater or in a special harness on land. This training is used to help you learn to maneuver in weightlessness. Normally we float in water. However, you can strap on weights that counterbalance the lifting capacity of your body underwater. Wearing these weights and being enclosed in a working space suit, you can float at a given depth without rising or sinking. Called neutral buoyancy, this condition simulates the microgravity you will experience in space. However, when you are in weightlessness training in water or on land wearing a "space suit," you will always have on an antinausea patch.

LOW AIR PRESSURE

One of the emergencies you may experience in space is loss of air pressure. This decreases the amount of oxygen in the air you are breathing, which leads to the medical condition called hypoxia (having insufficient oxygen in your blood). Decreased air pressure can occur for a variety of reasons, including air system hardware or software failure, puncture of the spacecraft by space debris, and leaks in your space suit. It is important for you to know the symptoms of hypoxia and how to respond to it before it becomes a potentially fatal problem.

You will probably have similar training in dealing with hypoxia that I had in the U.S. Navy's Flight Indoctrination Program in 1970 at the Naval Air Station in Pensacola, Florida: a session in a hypobaric chamber. Before we were taken flying, those of us in the program were put in the chamber, a school bus-sized room with seats facing inward. There were oxygen masks for everyone, and a table on which stood a glass of water. We were instructed on how to wear the oxygen mask. After everyone was then checked out, the door was closed and the air began to be pumped out of the room. The room began to cool and the glass on the table began to fill with bubbles, which rose to the surface as though the water in it was boiling. It wasn't boiling in the traditional sense of the word, but as the air pressure in the room decreased, the air trapped in the water was no longer being compressed as much by the surrounding atmosphere. The air was able to expand into bubbles and float upward, exactly as air does in boiling water or when you open a beer or a container of soda. We were instructed to take off our masks. The air was harder to breathe, meaning that inhaling didn't feel the same. I felt a shortness of breath and began coughing. We all had difficulty doing simple things, like clapping. We were then told to put our masks back on. One person began to pass out. The people sitting next to him helped him on with his mask. This hands-on training helped us better understand hypoxia so that we could deal with it effectively should the need arise.

Knowing the symptoms of hypoxia, including shortness of breath, sweating, rapid breaths, coughing, confusion, decreased coordination, and change in your skin color, can help you respond to it quickly in space by putting on an oxygen mask, turning on its air supply, and then sounding an alarm. This is one example of many contingencies you will be trained to deal with during your preflight training. Indeed, you will be taught the value of constant vigilance throughout your time in space.

SPACE SUITS

At some point, you will learn about space suits. You will be fitted for one, but it will not be like going to your favorite tailor. Space suits are worn during launch, during times outside of the space station (extravehicular activities), in emergencies, and when returning to Earth. The launch and reentry suits are relatively lightweight, enabling astronauts to walk on Earth in them as necessary.

Space suits designed at least in part for extravehicular activity are more complex and require more training for the user. Space suits and the accompanying undergarments are composed of several layers with different functions, including the maintenance of body temperature, the maintenance of air pressure so you can breathe comfortably, protection from the radiation in space and from the high-speed particles flying around out there, the removal of moisture that you give off from your body and breath, and the prevention of fire. Also, you will wear a diaper each time you wear a space suit since you may be cooped up in it for up to eight hours. Complicated, bulky, and uncomfortable, space suits are a major part of your wardrobe that you will have to tolerate.

You also need to know about the special peripheral equipment that space suits support. There are dozens of essential attachments to extra-vehicular space suits, such as communications equipment, sophisticated air-filtering and oxygen supply equipment, connections to tether you to space stations and spacecraft so that you don't drift away, and television cameras, to name a few. You will be taught how to don space suits under both normal and emergency conditions during the preflight training, and how to use the associated equipment.

The air you breathe will be different in spacecraft and space stations than it is in space suits. Except for trace gases like argon, neon, and helium, the composition and pressure of the air in space stations and space transport vehicles is the same as we breathe here on Earth, 78 percent nitrogen and 21 percent oxygen. The air you breathe in a space suit is likely to be much thinner (lower pressure) and have a different chemical composition than the air you are breathing right now. The reason all space suits to date have low air pressure is that the more air you put in a flexible container, whether a balloon or a space suit, the more the container stiffens (and in the case of the balloon, expands). If you were to pressurize any present space suit to 14.7 pounds of pressure per square inch, like the air we normally breathe, it would be so stiff that you wouldn't be able to move. Your arms would stick out at right angles from your body and your legs would form a V shape, which would not change until someone let the air out for you. Instead, space suits are pressurized to just 4.3 pounds per square inch (about 30 percent normal air pressure), and the air is pure oxygen.

This issue of higher-pressure space suits has an interesting history. In 1965, Soviet cosmonaut Alexi Leonov wore a suit initially pressurized to 5.87 pounds per square inch (about 40 percent normal air pressure) for the first-ever space walk. However, as the air was heated by his body and the Sun, the air pressure rose and the suit became so stiff, as described above, that after his 12-minute space walk he was unable to reenter the

Voshkod 2 spacecraft. This would, of course, have prevented him from returning to Earth. Fortunately, he was able to manually reduce the air pressure in his suit enough that he could squeeze back into the spacecraft.

Making the transition from normal air pressure to space suit pressure is nontrivial. You can't just put on the suit, purge the normal air, and fill it with 4.3 pounds per square inch of oxygen. Doing so would have the same effect as rising too quickly from a scuba dive. The air in your blood would form bubbles (like the air in the glass of water I saw in the hypobaric chamber), causing you to experience decompression sickness (the bends), which can cause various illnesses or death. Therefore, you will undergo a decompression period prior to donning a space suit, during which time your body will adjust to the lower air pressure.

The hassles associated with contemporary space suits could essentially all be short circuited if the air pressure in them could be adjusted to our normal air pressure without allowing the suits to balloon out and become stiff. Such designs have been under development for years. They are based on the principle of making the outer layer rigid, so the air can't push it out. A big challenge in making such suits work is designing flexible joints. These suits may be available when you go into space, making training on Earth and adjusting in space that much easier.

FLIGHT/SPACECRAFT/SURFACE SIMULATORS

Prior to liftoff, you will also participate in innumerable simulations of virtually every aspect of your experience in space, except possibly weightless sex. There will be high-tech mockups of launch/landing vehicles; all aspects of the orbiting hotel or other station that you will visit; the spacecraft taking you to the Moon or other destination body; landers for your destination; and, if you are going to colonize Mars, the habitat in which you will live.

In each of these simulators, you will undergo a variety of scenarios, both everyday and emergency, so that you will know what to expect and how to respond to myriad situations. For example, suppose your spacecraft is punctured by a meteorite and air starts leaking out. What do you do, and in what order do you do those things? Not only is it critical for you to know what to do, but also it is crucial that you know how to communicate effectively, how to work with others, and how to calm yourself in panic-inducing situations.

If you will be leaving Earth orbit, you will experience in underwater environments here what it feels like to walk on your destination world. Falling down and getting up again on the Moon are both very different than those experiences are on Earth or on any of the other destination worlds. Practicing before you leave will make your experience in space much more satisfying and comfortable. You will even take strolls on a surface similar to the world you plan to visit. For example, NASA uses various sites such as places on Antarctica, the area around Moses Lake (central Washington state), Black Point Lava Flow (near Flagstaff), Arizona, and various lava terrains on the Hawaiian Islands to simulate traveling on the Moon. It uses Devon Island, Nunavut, Canada (off the western coast of Greenland) to simulate Mars.

Fully garbed in a space suit suitable to your destination, you will practice doing the things that you will do off Earth. For example, you might trek several kilometers (miles) along mountain ranges or along canyon walls; ride a Moon buggy; explore caves that are likely to exist on various space destinations; and learn to collect and bag rocks and other samples from your destination world.

BATHROOM SKILLS

One skill in particular is highly valued in a spaceship. Going to the bathroom in space is something you will have to learn to do. The lack of gravity means that urine and feces don't go "down" up there. Therefore, you will be taught to urinate into a vacuum tube that will suck the fluid away, to be processed into drinking water. Similarly, you will learn to defecate on a toilet that holds you firmly in place and sucks away the feces deposited into it. Bear in mind that depending on your destination and the spacecraft that takes you there, the amount of privacy you have while going to the bathroom will vary.

Other basic bathroom skills, such as taking a shower, are fundamentally different in space than on Earth. Streams of water with which to shower are extremely impractical to maintain in microgravity, so you will clean yourself with towelettes. These are some of the small sacrifices you will make as a space traveler. People who venture into space are committed to exploration; creature comforts association with normal travel such as showers are a low priority.

Part II

ADJUSTING TO SPACE

———

5

LAUNCH!

—

Your launch into space will be one of the most exciting, satisfying, and scary experiences of your life. Given the considerable training you will have undergone, rehearsing every step of the process several times, you may find the preparations on launch day surprisingly straightforward. The evening before will likely be spent with family and friends, probably having an elaborate meal. After a good night's sleep, you will have a traditional prelaunch meal with your fellow space travelers. On your way to the bus that will take you to the launch vehicle, you can again see your friends and family, but for security reasons, you likely won't be able to touch them. That security, even in the best of circumstances, is to prevent your space suit from being damaged.

You will be wearing a space suit into orbit, which will protect you if air leaks out of your launch vehicle or any noxious gases leak into it from the fuel tanks and other sources. The suits used for takeoff and landing have air pressure inside them roughly equivalent to what you breathe in a jet flying at about 9,000 m (about 30,000 ft). Because the cabin in which you will sit during takeoff and landing has the same air pressure, the launch space suit will not balloon out when pressurized and will move as normal winter clothing does in everyday life.

Historically, the diapers worn during ascent and return to Earth rarely receive feces, since most astronauts use the bathroom for that purpose shortly before they put on their space suit. Likewise, careful intake of fluids limits the amount of urine absorbed by the diaper. In any event, modern diaper technology is so good that they can hold as much as 2 liters (l) (2 quarts) of fluid.

The space suit you wear out of the dressing area before launch is not fully configured with the hardware you will need. Some of that equipment is relatively delicate, so the less time you wear it, the less likely it is to be damaged. Also, some of the final equipment is bulky and/or uncomfortable, like the harness that will strap you into your seat (and possibly an emergency parachute), boots, gloves, and helmet. These items go on just before you enter the spacecraft.

SUBORBITAL FLIGHTS

Assuming that your trip will follow the mothership scenario described in chapter 2, your space plane will have been mated to its mothership before you board it. Recall that the mothership does the heavy lifting. While the equipment that will be attached to your suborbital space suit will be much less complex than that worn by travelers into orbit and beyond, you will likely undergo final dressing, such as attaching the helmet, communications equipment, and possibly an oxygen tank for emergencies, before you enter the space plane. You will be assisted in these steps by a ground crew.

Because experiencing weightlessness is a major feature of a suborbital flight, you will sit in seats set like normal aircraft seats from which you will be able to release yourself and float around the cabin while in space. Those seats will be much more substantial than normal seats, however, smoothly supporting your entire body—head to toe—for the acceleration you will undergo both on the way up and on the way down. Therefore, it is likely that the harness holding you in place will be similar to those worn by military pilots. It will go over both shoulders, rather than over just one, like a seat belt in a car. If your space plane is equipped with emergency ejection seats, then getting you settled in will be slightly more complicated, because you will also wear restraining straps that pull your legs firmly against the seat in case of ejection.

The ground crew will seat and secure you, and plug you in. They will check each system to which you are connected, including communication, oxygen, fluid intake, and, of course, your safety harness. Each passenger will get the same attention. When the ground crew leaves, securing the door behind them, you will be without a flight attendant.

The whine of enormously powerful jet engines will fill the cabin. You will likely have the option of listening to the radio chatter between ground control, which directs the movement of the mothership on the ground, and flight control, which controls all aspects of the flight, as well as pilots on the mothership, your pilots, and air traffic control. The latter group makes sure that no unauthorized aircraft are anywhere near you.

Taking off will feel like a high-powered version of a normal airliner taking off. A chase plane will take off with you. The job of its pilot is to watch for any signs of problems or potential problems during your ascent under the mothership: fuel leaks, damage to either your space plane or the mothership, and instabilities such as vibrations that either vehicle might undergo.

If all goes well, the mothership, accompanied by the chase plane, will ascend to the maximum altitude it is capable of. After an internal system check between it and your space plane, a final visual check by the chase plane pilot, and permission from all the related parties on the ground, your space plane will be unceremoniously dropped from the mothership. Seconds after you are released, the rocket at the rear of the space plane will ignite and the thrust will push you hard against the back of your seat. Your craft will be tilted upward and you will head toward the Karman line.

FLIGHTS TO ORBIT

All flights to orbiting space stations and beyond will start with flights into orbit well below the space stations, as discussed in chapter 1. After leaving your friends and family, you will be taken by bus to the launch pad. Leaving the bus, you will be lifted by an elevator or crane to the white room (so named because it is traditionally painted white), where astronauts finish getting into their space suits.

The equipment you put on in the white room will transform you from head to toe. The space suit you put on earlier will be filled with electronic equipment, emergency air tanks, water for you to drink, and, depending on the state of emergency technology, a parachute and survival equipment. A harness to hold you in securely will go around your suit. You will put on headgear that holds your communications equipment (headset and microphone). Your gloves and boots will be slid on and attached to the space suit. Finally, your helmet will be slid into place. It will likely have a retractable visor that can be pushed back over the top when a sealed suit is not required.

Your ride into space will be in a space capsule (plausibly a space plane) lifted by a rocket sitting vertically on the launch pad. Once fully suited, you will enter the spacecraft and take your seat. The support crew will connect you to the onboard communications, monitoring, and oxygen systems. After making sure that all systems are working, they will depart and the door will be closed and sealed. Although you will have rehearsed every part of this procedure several times, you will know that this is the real thing by the bustle and the feel of the rocket to which you're strapped coming to life. Your seat will be reclining so that you are lying on your back, with your legs and feet in a sitting position above your torso. In this configuration, you will just be pushed down in your seat during liftoff. Typical time strapped in before launch is about two hours.

The launch itself will begin with the countdown. NASA traditionally uses the letter T to indicate the planned launch time, which is why you hear "T minus 10, 9, 8. . . ." Your carrier may or may not use that terminology, but it is virtually certain that your launch will be preceded by "10, 9, 8, 7, 6, 5, 4, 3, 2, 1, liftoff!" During this countdown, the rocket engines will have already ignited. The rocket will be held onto the launch pad by clamps that are released when the rocket thrust has reached the required force. During this time, you are likely to feel considerable vibration.

You will feel many sensations on your trip into orbit. Once the rocket begins to climb, you will feel less or more vibration, depending on what types of rockets are being used to carry you into space. Liquid-propelled rockets are said to provide much smoother rides than solid booster rockets. Within seconds of liftoff, you will be pushed hard against your seat, feeling the force generated by the accelerating rocket, as discussed in chapter 2. You are taking the ride of your life. Depending on how many stages your rocket has, you will feel certain jolts as spent pieces of the rocket are released and later stages begin firing up. Within ten minutes of launch, the rockets will cease firing and you will be in low Earth orbit. If all goes well, within a very few minutes, booster rockets will ignite, causing your ship to spiral outward toward the space station, which will be either your destination or your first layover in space. You will arrive there about five hours after entering orbit. As noted in chapter 2, if anything goes wrong when you first get into orbit, arriving at the space station could take up to two days.

6

ADJUSTMENTS DURING
THE FIRST FEW DAYS

—

Although the human body did not evolve to live in space, its ability to adapt to a microgravity, high-radiation environment for weeks or months is impressive. In the rest of this section of the book we explore those adaptations, which begin immediately upon entering orbit above the Karman line on the way to a space station.

VISUAL AND MOTOR SKILLS

Ignoring the short bursts of acceleration that you undergo as your spacecraft moves into higher orbit or departs from Earth orbit, you will be weightless while traveling in space. You will know what this is like, as you trained for it (discussed in chapter 4). Many space travelers experience a variety of the neurological symptoms related to the loss of balance information from the inner ear. Besides space sickness, these include dizziness, vertigo (the sensation of spinning), and rapid, jerky eye motions. Even after these subside, your brain's adaptation to microgravity will show itself in a variety of physical and mental ways.

It is likely that you will have to relearn fine motor skills such as turning a knob or throwing a switch, even though you practiced these things underwater during your training. Networking with Earth will be a high priority with space travel companies, so that you can stay in close contact with friends and family. However, tweeting, blogging, and other computer-related activities that require careful use of your hands and fingers may be difficult for a while.

Perhaps one of the most confusing neurological problems you may experience early in your time in space is the inability to tell where different parts of your body are relative to each other and to other nearby objects. If it is convenient, try this right now: put your arms horizontally out to your sides. Close your eyes and then, one at a time, touch the pointer finger of each hand to the tip of your nose. Unless you have been drinking heavily before trying this, it should be no big deal because you "know" where your fingers are relative to your nose. This ability and the ability to physically move different parts of your body appropriately (hand to nose, for example) is called proprioception. I mentioned drinking here because heavy drinking compromises proprioception. That is why police will often test drivers they suspect to have been drinking by having them do what you just did, or having them try to walk on a straight, narrow line.

In the microgravity you will begin experiencing in orbit, your sense of proprioception will probably be temporarily compromised. When you reach for things, they may be closer or farther away than your brain thinks they are. Therefore, you may jab into them when you extend your arm too far or not reach them when you extend your arm too little. This applies to sensing both places on your body, like reaching down to take off your boots, and things nearby, like buttons to push.

A related adaptation that you will likely have to make early in your space travels is judging how much force to use in moving things. As you know, you and everything that you will come into contact with in orbit are weightless. However, everything out there still has mass, as discussed in chapter 1. You couldn't just lift a 208-l (55-gallon) drum filled with water while in microgravity and toss it like a feather. Rather, you must exert a force on anything you want to move. The difference between here on Earth and out in space is that once you have forced an object to move up there, it will keep on moving, rather than "falling" back down under the force of gravity. Even knowing this, you may not initially move things appropriately. Indeed, it will take hours of practice grabbing and moving things before you can do this reliably and well.

BODY FLUID REDISTRIBUTION

Bodies need to adjust to microgravity. Our bodies evolved to take advantage of the force of gravity where possible. In particular, blood leaving the heart, going "downward," namely into our legs, is assisted in its flow by the pull of gravity when you are standing or sitting. This enables the heart to do less work than if it had to push the blood in all directions through the body with equal force. In the microgravity of space, that downward pull from gravity vanishes. The heart doesn't receive much information about lower blood flow, so the fact that the legs are not getting their usual supply is not compensated for by the heart beating harder and faster to push the blood down there. With less blood flowing in them, your legs will immediately start to become more and more spindly. They will revert to their normal size and structure after you return to Earth.

The blood that normally would go to your legs is still in your body, of course, but now it is concentrated in your head, arms, and torso. As a result, these regions will quickly begin swelling. Your face will look puffy, which is one of the reasons that most astronauts are not photographed early in their missions.[1] Related signs and symptoms include increased sinus congestion, severe headaches, oily skin, and distended (swollen) jugular veins.

The extra blood now flowing in your brain has the potential to cause serious damage, which is why the brain immediately starts sending hormonal signals to the kidneys and related organs to remove the excess fluid from your body. You will therefore urinate more than usual until your brain feels that the amount of blood flowing through it is again normal. The new equilibrium in your circulatory system will also include a decrease in blood pressure. While this is conceptually a good thing, keep in mind that if you have to suddenly do some hard physical work, your heart will not be working very hard and you may momentarily feel faint or dizzy, and your vision may blur. These effects will pass when your heart beats hard and fast enough to supply the oxygen that your cells need during this time.

NUTRITION AND DIGESTION

An activity that we take for granted here on Earth, at least until it stops working correctly, is our digestion of food. This is a complicated process that takes many steps and several hours from the time we eat something until any parts of the food we can't use are eliminated from our bodies. Because our bodies evolved in adaptation to the Earth's gravitational force, the fluid balance in our bodies, the bacteria in our small intestines, and our normal motion enable this processing to occur. Since virtually all of these factors change when we are in microgravity, they will have an impact on your eating and processing food.

Eating well is often one of the pleasures of life. It is, however, one that you will have to face losing while you are in space. The problems are likely to start as soon as you enter microgravity. It would not be uncommon if your bowels stopped functioning on the first day in space. This is called ileus. The causes of it are not completely clear, but microgravity apparently plays a variety of roles, including decreasing the gravitational force on the food you are digesting, changing the normal flow of this material through your bowels, changing the bacterial concentrations in your intestines, and decreasing the muscle activity of peristalsis that drives materials through the intestines. This shutting down of your bowels also occurs on Earth as a consequence of a variety of surgical procedures, for instance.

It is essential that your digestive tract work, at least to some degree, before you start eating in space. The consequence of not being able to digest food is that you will be unable to consume it during this period, either. Should you eat while your bowels are not functioning, you might vomit and possibly aspirate (breathe in) some of this material, causing you to choke. Fortunately, it is easy for a trained observer to determine if you are suffering from ileus by listening to your bowel sounds. It is worth checking before you learn of the problem "the hard way." Digestion usually begins within 48 hours of the time you enter microgravity, so be patient. It is not a good idea to rush this process, since taking laxatives can lead to diarrhea, which you really don't want to have in space. Even when you are not eating solid food, it is essential for you to drink water.

You will have a wide variety of foods to choose from when you plan the meals for your journey. Even today, astronauts have over 400 food and drink choices. However, you are likely to find that the foods you enjoy on Earth taste much blander in space, which is why McDonalds in space will have to change the spices in its menu.

Your caloric needs are essentially the same on Earth and in space. Even after the initial digestion problems go away, most astronauts eat less than they should. This is an especially serious problem for people on long-term space missions, as the consequences are cumulative. Part of the loss of appetite occurs because the food tastes bland, part because digestion takes longer and so they feel full longer, and part because the redistribution of fluids, discussed earlier in this chapter, makes eating much less enjoyable.

Following your instincts and eating less when you are in space will create more problems than it solves. Insufficient nutrition leads to a variety of illnesses and weakness, including bone loss, overall weight loss, muscle mass loss, and vitamin and mineral deficiencies. Being malnourished will negatively affect both your enjoyment of the trip and your readjustment when you get back to Earth. I will have more to say about that readjustment in chapter 10. While a variety of medical interventions are being explored to help you maintain a healthy appetite, one thing that you can do is to order much more heavily spiced foods for your trip to help make eating more appealing.

Historically, drinking coffee and other fluids in space has been done through a straw from a sealed flexible packet. However, the highly unintuitive behavior of fluids in microgravity will actually allow you to drink through a specially designed cup (developed in 2013), which can be a more satisfying experience, since you can smell the coffee or other fluid in the cup as well as taste it. The lip of the cup designed for sipping fluids in space isn't perfectly round. Rather, it has one side where two flat edges come together in a V shape. Fluid is gently squirted into the cup, where it all stays together, clinging to the side of the cup. Unless you force it out, it will stay as one fluid clump. Remember, because it is in microgravity, there is no "down," so if you tilt the cup in any direction, the fluid just sits there. The secret to drinking coffee or other fluids from a cup in space is to put your lips to the V and sip. This activity creates a partial vacuum, like a vacuum cleaner, causing the fluid to flow toward you so you can drink it. This is one of many examples of where common sense, intuition, and experience don't work in space!

GETTING LONGER

If you ever wished you were taller, space travel can help you, at least temporarily. While bones maintain your body's overall shape, the way that they are connected also gives you flexibility. From head to hips, your body is supple in large measure because your spine is composed of 33 bones (vertebrae). Twenty-four of them are separated by cartilage and held together by muscles, ligaments, and tendons. (The other nine are fused together.) The muscles allow the separate vertebrae to pivot front to back and sideways relative to each other, enabling you to bend and twist. While you are in microgravity, however, your spine is not compressed by gravity; therefore, the support system of muscles, ligaments, and tendons relaxes. As a result, your back will stretch out or decompress by as much as 5 centimeters (cm) (an inch or two) in just the first few days of your trip. Indeed, astronaut Jerry Linenger reported growing over 5 cm (2 in) in two days!

There is a second way you will get longer. Tonight, when you are lying in bed, think about your feet. How are they pointed? You will find that instead of sticking out perpendicular to your legs, as they do during the day, your feet will be pointing almost directly away from your head; they will be nearly parallel to your legs. In microgravity your feet will tend to point down all the time. This is called foot-drop posture, and it contributes to muscle loss in your legs. The good news is that cycling and treadmill workouts both enable you to use your feet in their normal position, which helps to slow the loss of muscle in them. All space stations have such exercise equipment. Their value in helping maintain astronaut health is so significant that all future space venues, as well as spacecraft taking you to other worlds and hotels on the Moon and Mars, will have a variety of exercise machines.

7

LONG-TERM PHYSICAL ADJUSTMENTS TO SPACE

———

To help us understand the physiological effects of long-term space-flight, astronaut Scott Kelly and cosmonaut Mikhail Korniyenko both spent nearly a year in the *International Space Station*, beginning March 27, 2015, and returning to Earth on March 1, 2016. Scott and his brother Mark Kelly are identical twins. Mark, husband of former U.S. Representative Gabby Giffords, has also flown in the *International Space Station*, but for a shorter time than Scott. What makes Scott's year in orbit especially interesting medically is that not only will physicians be able to see how he has changed over a year, but they will also be able to compare him to his brother both physically and in terms of any genetic changes that he might have experienced during that time. This is still a work in progress.

MEDICAL ADJUSTMENTS TO MICROGRAVITY

Your body will also start undergoing several long-term adjustments once you enter microgravity. These changes, such as bone and muscle loss, can have serious consequences for you once you get back on Earth. For example, less dense bones are more prone to fracture; likewise, less muscle mass can lead to injuries and falling if you try to do strenuous physical things immediately after you return. Therefore, exercising and eating appropriately in space are essential to minimizing these and other physical effects of microgravity.

Bones

We maintain our overall shapes because we have bony skeletons. One of the major problems that occur in microgravity, made worse by poor nutrition, is the degradation and loss of bone. (For example, if you don't consume enough vitamin D3, which is necessary to maintain bone health, the loss of bone mass will be accelerated.) Furthermore, we are able to move because our muscles can use the bones as levers and fulcrums. The combination of rigid parts (bones) and engines (muscles) enables our bodies to change position and location, while protecting all the delicate parts inside our ribcages.

Bone has other roles besides structure and motion. While some bone is composed of just rigid material that gives the body shape, most bones have space inside them filled with spongy bone marrow in which the different types of cells for your blood are manufactured.[1] Bones change significantly in space. Two major chemicals they lose are calcium phosphate, which gives bones their rigidity, and the element calcium, which is stored in bones for use throughout the body as necessary. For example, the otoliths in your ears' vestibular apparatus are made of calcium carbonate compounds. Loss of bone minerals (osteoporosis) is a double whammy because it makes your bones more brittle and also starves other parts of your body of the calcium compounds they need. Doctors and scientists are studying this bone mineral loss. They do know that the loss is related in part to your reduced weight on other worlds and your weightlessness in space. In either case, your body does not require as strong a support structure as it does when you are on Earth.

Interestingly, the alterations apply primarily to the bones of the legs, pelvis, and lower back, which lose 20 times as much mass as the bones in the upper body. The most vulnerable parts of your skeletal system are the upper part of your femurs (long leg bones between your hips and knees), which initially lose up to 1.6 percent of their minerals per month; your pelvis, which initially loses about 1.4 percent per month, and your lower spine, which initially loses about 1.1 percent per month. Your body's overall average initial bone mineral loss is 0.35 percent per month. Put another way, every four days in space during the initial months, you will lose about 0.002 pound of bone mineral. These numbers are much higher than the amount of bone lost by postmenopausal women on Earth.

Fortunately, these numbers constitute the *initial* mass loss rates. You will not eventually become a mushy blob in space. It remains to be seen if Scott Kelly and Mikhail Korniyenko lived in space long enough to establish a final "steady state" for bone mineral concentrations. All the other (shorter-term) astronauts have continued to lose bone minerals to some extent throughout their missions. We can only say, extrapolating from their experiences, that if you travel out there long enough, you can expect your bones to stabilize at a much lower concentration of the calcium minerals than you have today.

The consequences of bone mineral loss are extremely serious. Most important for you during your trip is the danger of breaking a bone either by accident or by exerting force on your skeleton that would normally not pose any problem here on Earth. In other words, if after a six-month voyage to Mars you lifted something that weighed 22 kilograms (kg) (50 pounds) there, you would be at risk of breaking a bone that would easily have been able to sustain the stress of lifting that weight back on Earth.

Space medicine has yet to deal with broken human bones, so physicians have yet to discover how best to set them in microgravity or other low-gravity situations. Healing processes in space typically take much longer than they do on Earth, and a bone that is broken and heals in space is expected to be much weaker than a bone that heals on Earth. You may wonder why space travelers with broken bones would not just be given calcium-rich foods or supplements to aid in the healing process. It makes sense at first glance, but actually that could create even more problems than it might solve.

The human body is so complex that making one change to it often leads to a variety of other changes. Because space travelers are continually flushing calcium out of their bodies from their bones, their blood and urine are continually rich in calcium-based minerals. This excess calcium in urine is called hypercalciuria. Therefore, if you broke a bone and were given more calcium, you would not absorb all of that mineral in the bone-healing process. Most of it would be flushed out of your system, increasing your hypercalciuria. The major immediate consequence of this condition is the formation of kidney stones, which often cause excruciating pain and can be accompanied by nausea and vomiting. On Earth, kidney stones often pass without surgical intervention. It remains to be seen if the same is true in space. Consuming large quantities of water helps in this process, but you won't have a lot of water in space.

The space medicine community expects that bone loss in space will lead to a significant increase in the likelihood that you will experience osteoporosis as you get older back on Earth. Fortunately, a considerable amount of work is already being done to prevent or alleviate the symptoms of this subtle, yet dangerous condition. But you should know that your trip puts you at increased risk for it.

Teeth

Teeth decay on Earth relatively slowly, and once a cavity is identified, either by the pain it causes or during a dental examination, the repair process is relatively rapid and relatively painless, although expensive. Cavities are caused by naturally occurring bacteria in our mouths that eat leftover food and excrete acid, which in turn removes the outer layers of unprotected teeth. Ideally, we return the lost minerals to our teeth as fast as they are removed. When this does not occur, decay sets in. Fluoride treatments harden the teeth, making them more resistant to decay.

Teeth in space are at increased risk of decay because decay-producing bacteria reproduce 40 to 50 times faster in microgravity than on Earth. In 1978, cosmonaut Yuri Romanenko, orbiting in the *Salyut 6* space station, had a cavity exposing the tooth nerve, which caused him excruciating pain that he had to endure for two weeks, until the end of the mission. Besides natural tooth decay, any number of incidents can cause damage to teeth, such as being struck by massive floating objects in microgravity or even natural grinding of teeth or breaking a tooth while eating.

Thus, much thought has gone into how to care for tooth emergencies in space. The issue of potential dental problems both in space and in long-term isolation on Earth, such as on submarines, has prompted the ongoing development of prophylactic and restorative measures for dental events in those environments. These are intended to relieve pain, prevent secondary infection, and protect teeth until they can be dealt with by professional dentists.

Muscles

The loss of tensile strength in your bones is paralleled by the loss or atrophy of muscle mass connected to your skeleton, along with the conversion of other skeletal muscles from one type to another. The first effect is easy to understand. It basically amounts to "use it or lose it." Atrophy of muscle tissue typically begins within five days of the time you enter microgravity.

The second effect requires some explanation. You have a variety of muscle types in your body. One group, called slow-twitch or red muscles, are rich in blood, have the ability to maintain energy supplies for a long time, and are therefore slow to fatigue. The energy is most commonly stored in a molecule called adenosine triphosphate or ATP. The slow-twitch muscles provide you with the capacity for long-distance running, bicycling, swimming, and other aerobic activities. As the slow-twitch muscles become idle in space, they frequently begin to ache, causing back pain in a fair number of space travelers.

Another group, called fast-twitch or white muscles, don't have large stores of energy, nor can they resupply themselves quickly. They are the muscles that provide power when you have short bursts of activity, such as you would need when sprinting or lifting large amounts of weight. Studies of astronauts reveal that red muscle tissue converts to white muscle tissue in microgravity. Therefore, without conditioning while you are en route, you will find that your walking, climbing, and running endurance on other worlds will be far below what you used to be able to do on Earth. While the increase in white muscle tissue will occur, its positive effects on your ability to lift, for example, will be diminished by the fact that you are simultaneously losing overall muscle mass.

Part of the problem of losing muscle is probably poor nutrition. Overcoming this and the possible effects of various medical interventions on preserving muscle are still being studied. Since you will be losing the muscle useful for aerobic (endurance) activities, you might expect that doing aerobic exercises in space, such as treadmills, rowing, or bicycling, would help maintain your red muscle mass. So far, however, the experiments in space using such devices have proven only modestly effective in minimizing the conversion process, and not terribly effective in slowing muscle atrophy.

Speaking of muscles, don't be surprised if you or others become cross-eyed. Called dysconjugate gaze, this is a common occurrence that can persist even for some time after you have returned to Earth.

Medicine

Space medicine is now a full-time profession. Along with the physical changes to the human body, some of which are described above, are changes in our response to medicines. The microgravity of space causes drugs to be absorbed at different rates than they are here on Earth. Therefore, doses of all the medicines you take have to be reassessed for use in space. This is one of the many issues under investigation in the burgeoning field of space medicine. Considering the variety of drugs that are used and available both over the counter and by prescription, and the relatively small number of people who will precede you into space, chances are that you will have to take meds up there whose microgravity doses are not yet known. Consider your response to them as a contribution to the body of medical knowledge so that people who take them in space after you will know better what doses they need.

Adjusting medicine doses in space is made even more complicated by the fact that many medicines lose potency much more rapidly in space than they do on Earth. Put another way, medicines in space expire well before the dates on their labels. The most likely reason for this, yet to be confirmed, is that the radiation to which they are exposed modifies their chemistries. That radiation, discussed in chapter 1, is normally prevented from reaching the Earth's surface by our planet's atmosphere. Since the chemistry of each medicine is very specific in determining how it affects our bodies, changes in their chemistries lead to loss of effectiveness. At the very least, dosages will have to be revised as a function of how long each medicine has been in space. This is one of the many aspects of space medicine that will evolve significantly in upcoming years.

LIGHTING, CIRCADIAN, AND OTHER RHYTHMS

Our bodies are controlled by clocks "ticking away" inside us. Your brain, for example, has oscillators that create brain waves that pulse through it at different rates to help it organize and regulate ongoing activities. Called alpha, beta, gamma, delta, and theta waves, they typically oscillate at frequencies between a few thousandths of a second and about 10 hertz (10 times a second). Many of our activities, such as waking and sleeping, metabolism, and body temperature, are called circadian (meaning "about a day") rhythms that are regulated by the body's biological clock, located in your brain in an area called the suprachiasmatic nucleus. Cycling about once every 24 hours, it regulates when you feel the need to sleep and wake, when you are hungry, and other relatively long-term activities.

In order for us to function on Earth with its 24-hour day, our biological clocks are reset every day. This is done most effectively by changes in the amount of light we receive. For example, the early morning sunlight helps reset your biological clock, which is physically located very close to the optic nerve. The light helps your body turn off the sleep-inducing chemicals, such as melatonin, that your brain secretes. Indeed, during parts of the year when the Sun isn't up at rising time, it is often harder to start the day since your biological clock doesn't get the powerful resetting that light can create. If you ran on a light-dark (day-night) cycle that was significantly different than 24 hours long, you would not be able to re-sync your biological clock each "morning," which would lead to both severe physical and severe emotional problems.

Resetting biological clocks each day is essential. Artificial lighting and active climate cycles are needed in environs where lighting and climate cycles don't normally occur, such as in submarines, in surface ships where sailors don't get light from outside, and above the Arctic Circle and below the Antarctic Circle (regions with days or months of continuous darkness and of continuous light).

With the exception of the surface of Mars, time cycles in space are not remotely close to our normal daily cycle of 24 hours, as you can see in table 7.1. For example, the day-night cycle in low Earth orbit is about 90 minutes long, as noted earlier. During that time you will have 45 minutes of daylight and intense heating of the spacecraft followed by 45 minutes of darkness and intense cold. (While the temperature outside the spacecraft cycles dramatically every 90 minutes, it is well regulated inside all spacecraft.) If you had to live and sleep by such a short cycle of light and dark, you would be unable to function well or to enjoy yourself. Therefore, life in low-Earth orbit is regulated by artificial cycles of active lighting in your ship or space station.

Interplanetary space lacks a natural day-night cycle. When you look out a porthole, the sky all around your ship will always appear pitch black because very little gas exists in space to scatter sunlight the way our atmosphere does to create daylight. On any of the worlds you visit except Mars, even when the Sun is "up," no bright sky will supply the amount of sunlight you experience on Earth because there is no atmosphere to scatter sunlight and thereby glow as our sky does. Twenty-four-hour time cycles useful to humans will have to be artificially imposed on the habitats set up on other worlds besides Mars. Likewise, on trips away from the Earth-Moon system, 24-hour light and temperature cycles will be programmed on your spaceship.

TABLE 7.1 Day and night periods of some nearby objects

OBJECT	LENGTH OF DAY-NIGHT CYCLE
our Moon	29½ Earth days
Mars	24 hours, 40 minutes
Phobos	7 hours, 39 minutes
Deimos	30 hours, 20 minutes
typical Comet	a few hours
asteroid Eros	5 hours, 16 minutes
low Earth orbit	90 minutes

Enough people have flown in space for physicians and psychologists studying space medicine to note a high frequency of complaints about sleep disturbances and time disorientation, as are also noted in Antarctica and in submarines. Let's consider the crucial issue of sleep.

SLEEP DISTURBANCES

Noise

Noises and vibrations often spoil a good night's sleep. Noise is basically sound that carries, from your point of view, much less information than its volume, pitch, or duration warrant. The sound of a rocket roaring underneath you coincides with the acceleration that tells you the ship is changing speed, but other than that, the sound of a rocket is usually just an irritation. It is interesting that sounds usually perceived as noise can be very welcome. Imagine that you are traveling to the Moon and have been told that there may be a problem with the ignition of the rocket that will slow your ship down and bring it into lunar orbit. Nothing in the universe would please you more than to hear the deep-throated roar of the engine at the appropriate time. It would be "music to your ears."

Noise (sound with little information content) is evaluated in a variety of ways:

- frequency or pitch (a high-pitched squeal versus a deep bass roar, for example)
- loudness
- whether it is continuous or intermittent
- whether it is expected or unexpected
- whether it is necessary or unnecessary
- whether it is related to a life-sustaining activity or not
- whether everyone agrees that it is noise

The latter point refers, for example, to the subjective feeling that certain music is good or bad. Bad music is just "noise" to the listener who doesn't like it.

Because sound is so important in our lives, its intensity is quantified in units called decibels, denoted db. The decibel is a mathematically defined measure of the amount of sound energy passing any point. For our purposes, it is just useful to know some of the values, which are listed in table 7.2.

Let us look at the consequences of extremely loud noise. If you were to stand 15 meters (50 feet) from a rocket lifting off the Earth, you would hear over 200 db. As you can see from the list, this is lethal. Long-term exposure levels above 60 db are considered harmful to your hearing; persistent noises above 85 db will definitely lead to hearing loss. The noise levels in the *International Space Station* are often above the 35 db level recommended for sleeping and above the 50 db limit that was intended for the craft in general. To counteract the effects of constant noise, people living in it frequently wear earplugs or noise-canceling headphones; this effectively reduces the physical and mental strain that the noise generates. Indeed, the *Mir* (Russian for "peace") space station was so noisy that some cosmonauts who spent months in it have suffered permanent hearing loss.

Noise is often distracting: it significantly affects our quality of life. Job performance often degenerates in noisy environs when compared to the same work done under quiet conditions. Noise also has the effect of making people tire more easily, which decreases job efficiency, stamina, and enjoyment. If your sleeping environment is noisy, the duration and quality

TABLE 7.2 Effects of noise at various volumes

LOUDNESS (DECIBELS)	TYPICAL EXAMPLE OR THE EFFECT OF THAT LOUDNESS
20	whisper
60	typical restaurant noise
80	standing on overpass above busy interstate
140	causes physical pain
200	causes death

of your sleep will both be lower. And, of course, noise that prevents you from hearing instructions clearly is a potentially serious issue when you are involved in life or death activities in space.

Even when you are coasting (no rockets firing), as is likely to occur for most of the time you are in space, you will face noise with a variety of characteristics. What makes spacecraft noisy? Fans, pumps, motors, water pipes, people moving about and talking, humming transformers and other electronics, electromechanical valves opening and closing, and, of course, music leaking from headsets, among other things.

One way to decrease the effect of all this background noise is to lower the cabin air pressure. The lower the pressure, the harder it is for sound to travel. The problem is that lower air pressure also makes it harder to be heard when you talk. Therefore, you would have to speak louder, which is likely to give you a sore throat if you speak a lot. Another way to minimize the physical and psychological effects of noise is to wear soft earplugs or noise-canceling earphones, both of which are available and are also quite handy to bring on airplane flights today. The problem with basic noise-reduction technology in space is that you become more isolated from the people around you, which, as we will discuss further in chapter 8 on psychological issues, is not a good thing. Happily, it is possible to use wireless, noise-reducing headsets with microphones that enable you to converse comfortably with others.

Vibrations

Whereas noise could be defined as undesirable vibrations in the air detected by our ears, vibrations of the solid and liquid elements of your surroundings in space are also undesirable. As with sound, vibration has a variety of "dimensions," including:

- frequency
- intensity
- duration
- whether it is continuous, occurs on and off at predictable intervals, or is sporadic
- direction, such as up and down, twisting, or variable

Spacecraft vibrate powerfully as they lift off the Earth. Indeed, this event should produce the most intense vibrations that you will feel in space. However, even when rocket engines are turned off, vibrations remain from many of the sources that also generate noise.

Vibrations at different frequencies affect us in different ways. Some frequencies cause you to feel that your skin is vibrating, while other, usually lower frequencies can make your insides feel as though they are jumping around. Some of the frequencies and the parts of your body they affect most are listed in table 7.3.

TABLE 7.3 Vibration frequencies that affect various parts of the human body

BODY COMPONENT	RESONANT FREQUENCY (hertz, or cycles per second [Hz])
whole body, standing erect	6 & 11–12
whole body, standing relaxed	4–5
whole body (transverse)	2
whole body (sitting)	5–6
head	20–30
head, sitting	2–8
eyeball	40–60
eardrum	1,000
head/shoulder, standing	5 & 12
head/shoulder, seated	4–5
shoulder/head, transverse rib	2–3
main torso	3–5
shoulder, standing	4–6
shoulder, seated	4
limb motion	3–4
hand	1–3

(continued)

TABLE 7.3 *(continued)*

BODY COMPONENT	RESONANT FREQUENCY (hertz, or cycles per second [Hz])
thorax	3.5
chest wall	60
anterior chest	7–11
spinal column	8
abdominal mass	4–8
abdominal wall	5–8
abdominal viscera	3–3.5
pelvic area, semisupine	8
hip, standing	4
hip, sitting	2–8
foot, seated person	>10

Reference: NASA Man-Systems Integration Standards Figure 5.5.2.3.1–1

The vibrations passing into your body can cause you to develop both functional and physical problems. Functional problems interfere with activities, as listed with the relevant frequencies in table 7.4.

These problems relate to how the mind and body interact. They are compounded by a variety of physical discomforts also caused by vibrations. Many of these are listed in table 7.5.

When you are strapped down in microgravity so that you can feel the habitat around you, these potential irritations are more likely to occur. You can minimize the dangers and discomforts by floating freely (when possible).

Sleep Disorders and Dreaming

Sleeping well is at least as important in space as it is on Earth. We have explored how your body and mind will undergo a variety of physical changes

TABLE 7.4 Activities affected by vibrations of various frequencies

ACTIVITY AFFECTED BY VIBRATIONS	FREQUENCY RANGE OF VIBRATIONS (HZ)
equilibrium (balance)	30–300
tactile sense	30–300
speech	1–20
head movement	6–8
reading (texts)	1–50
tracking	1–30
reading errors (instruments)	5.6–11.2
manual tracking	3–8
depth perception	25–40, 60–40
hand grasping handle	200–240
visual task	9–50

Reference: NASA Man-Systems Integration Standards Figure 5.5.2.3.2–1

when you leave the Earth. Without quality sleep, you will have great difficulty coping with these changes and will be less alert in an environment that requires constant vigilance. During sleep our brains repair and reset themselves. When the day-night and other major cycles in space fall outside the bounds that your circadian rhythms can reset, you will begin to suffer. During these periods, sleep is not as deep as it needs to be, and you will not spend the proper amount of time in each of the five stages of sleep, including the period during which you have vivid dreams.

These sleep problems are often made worse by changes in other parts of a space traveler's brain, especially the hypothalamus and pituitary gland. These areas interact by a complicated chemical feedback mechanism responsible for normal growth in childhood via the secretion of human growth hormone, among other things. In microgravity a chemical imbalance often occurs in this glandular system, causing renewed secretion of human growth hormone. This leads to a cascade of potentially debilitating chemical changes in the brain that affect sleep. They also keep people

TABLE 7.5 Physical problems created by vibrations at various frequencies

SYMPTOM FREQUENCY	FREQUENCY RANGE OF VIBRATIONS (HZ)
motion sickness	0.1–0.63
abdominal pains	3–10
chest pain	3–9
general discomfort	1–50
skeletal or muscular discomfort	3–8
headache	13–20
lower jaw symptoms (clenching, teeth chattering)	6–8
influence on speech	13–20
"lump in throat"	12–16
urge to urinate	10–18
influence on breathing	4–8
muscle contraction	4–9
testicular pain	10
dyspnea (discomfort in breathing)	1–4

feeling "low" both physically and emotionally, impairing their ability to make sound judgments, causing them to respond oddly to stress, and causing abnormal behaviors.

Sleep disturbances are among the most common disorders reported in space, with astronauts and cosmonauts typically getting two hours less sleep than they need each "day" unless they take medication. The need for quality sleep of the proper duration is so essential that about 75 percent of astronauts use sleeping pills. These account for some 45 percent of all medicines taken in space today. However, none of them has proven to be entirely satisfactory in the depth and duration of sleep they induce or in solving the problem of side effects. New medicines that have recently been approved may help alleviate these troubles.

Once your sleep cycle is disrupted, the result is usually too little sleep. Unfortunately, the obvious remedy of taking caffeine and other uppers to help stay awake has the effect of compounding all the cognitive problems that you are developing, making you more irritable, less rational, and less able to reason properly. Of course, these problems can have grave consequences during space travel.

Besides having to cope with sleep disorders of your own, you are likely to encounter situations where others affect your sleep or vice versa. Someone who wakes up screaming from a night terror or nightmare is likely to awaken others, as are insomniacs who are up when others are trying to sleep. Also, people who snore on Earth are prone to snoring in space. As the body of knowledge concerning sleep, especially how it is different in space, grows, improved methods of assuring quality sleep should be available.

EXPOSURE TO RADIATION IN SPACE

Above Earth's atmosphere, space is awash in harmful electromagnetic radiation. Ultraviolet, X-ray, and gamma ray photons have sufficiently high energies to directly damage DNA and other biological tissue, as discussed in chapter 1. While space vehicles provide some protection, you will have a higher level of exposure to all these photons than you have on Earth. This is especially true when you are walking on another world, since space suits normally provide less protection than do spacecraft or space habitats.

Cosmic Rays

The damage that high-energy photons can cause is compounded by the damage caused by cosmic rays, high-speed atoms that permeate space. First discovered in 1912 by Austrian physicist Victor Hess (1883–1964), they were given the name "cosmic rays" in 1926 by the American physicist Robert A. Millikan (1868–1953), before they were known to be particles. They are often called galactic cosmic rays, since they come from outside our solar system. As sometimes happens, the misnomer "rays" stuck, a source of great confusion among physics students and others. Observations reveal that 85 percent of galactic cosmic rays are protons, 14 percent are helium nuclei (also called alpha particles), and the remaining 1 percent are composed of virtually all other types of naturally forming atomic nuclei, as well as electrons.

The speed and mass of a cosmic ray determines how much energy it packs, and hence its effects when it interacts with other things in the universe. Such energy, due to an object's motion, is called kinetic energy. Unlike billiard balls, however, atoms, ions, and electrons are not solid. They don't just carom off the particles they encounter, like the cue ball off the six ball. Like photons, galactic cosmic rays have both particle and wave properties that complicate their interactions, as introduced in chapter 1.

Cosmic rays from outside the solar system are among the most powerful projectiles passing through nearby space. Shining stars, exploding stars, colliding stars, and colliding stellar remnants such as neutron stars and black holes, among other sources yet to be determined, create them. These "rays" are typically much more powerful than the particles of the solar wind, the gas that the Sun continually emits. The solar wind particles are therefore not considered to be cosmic rays. However, the Sun does emit bursts of energetic particles that are also classified as cosmic rays and are exceptionally powerful and dangerous. Earth's magnetic field captures solar wind and some solar cosmic rays. Sometimes these particles will leak out of the field, especially near the poles, and cause the air to glow, which we call aurorae.

Most galactic cosmic rays are so energetic that they pass through the Van Allen belts. Fortunately, our atmosphere prevents all but the most energetic ones from reaching Earth's surface. This stoppage often occurs when the cosmic rays smash into gases in the air, causing the gases to break up into smaller particles, which are sent Earthward at high speeds. Many of these particles soon collide with other particles in the air, creating a cascade of particles. The transfer of energy from high-energy particles in space, formally called primary cosmic rays, to particles in our atmosphere creates a secondary cosmic ray shower (figure 7.1). This process often continues until some of the particles strike the Earth and things on it, like us.[2] Because each collision takes some kinetic energy away from the incoming particle, the energies of secondary cosmic rays at the Earth's surface are much lower than the energy of the primary cosmic rays that started the shower.

Let's look at the effects of high-energy particles on life in space. These particles are often powerful enough to penetrate the few centimeters (inches) of shielding[3] on the *International Space Station* or other craft and habitats you may visit in space. As for human damage, some 5,000 of these particles will be going through your body every second that you are in a space vehicle or taking a space walk. Most of them will damage cells as they go. The more massive cosmic ray particles, such as iron and nickel, will do much more damage than the lowest-mass hydrogen nuclei and electrons. Indeed, the highest-speed iron nuclei galactic cosmic rays carry as much punch as a baseball thrown at 100 km/h (60 mph). All these impacts will injure cells, causing some of them to die and damaging

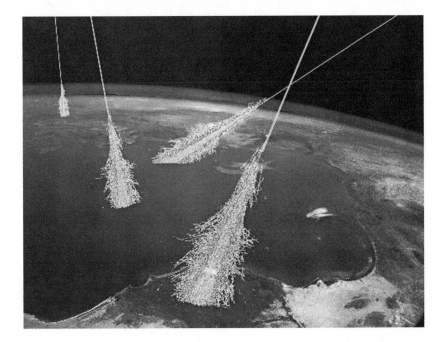

FIGURE 7.1

Drawing of four cosmic ray showers, in which a high-energy cosmic ray from space causes a series of particles in the air to move downward, colliding with other particles.

Simon Swordy (U. Chicago), NASA

others, which your body will try to repair or replace. New studies suggest that astronauts exposed to galactic cosmic rays for sufficient lengths of time are at increased risk for developing dementia.[4]

Radiation in Different Locations

We humans have been exposed to dangerous doses of radiation (including primary cosmic rays when we are high in the atmosphere or in space) for only a little more than a century. Severe exposure has been limited to relatively few people. Besides astronauts, they include those who were exposed to nuclear explosions (in both peacetime and wartime), nuclear accidents related to nuclear fuel, the refinement of radioactive ores, the Chernobyl and Fukushima nuclear power plant disasters, and laboratory radiation (especially before the dangers of radioactive materials were understood). Consequently, science and medicine's understanding of radiation effects on humans is as yet incomplete.

It *has* been established that various systems and organs in our bodies differ in their sensitivities to any radiation entering us from Earth or from space. Listed below are ten human organ groups in order from most to least radiation sensitive:

1. blood-forming organs, including lymph nodes, the thymus and spleen, and bone marrow
2. reproductive organs
3. digestive organs
4. the circulatory system
5. the skin
6. bones
7. the respiratory system
8. the urinary system
9. muscles and connective tissue
10. the nervous system

It is interesting that our nervous system—the brain, spinal cord, and peripheral nerves—is relatively insensitive, since it is arguably the most complex system in our bodies. This is true in part because the rate of cell division and cell replacement in the nervous system is relatively low (at least in adults), so the most vulnerable stages of the cell reproduction cycle are less likely to be present when the radiation penetrates the body than they are in other organs. Since the juvenile nervous system changes much more than an adult's, it is likely that children will be more sensitive to radiation in space than adults.

Consider next the effects of radiation on human bodies. Listed here are the top ten effects of short-term high doses of radiation.

1. reddened skin (erythyma)
2. fatigue
3. diarrhea (due to the breakdown of the linings of the stomach and intestines)
4. nausea
5. vomiting
6. skin blisters
7. dehydration
8. hair loss
9. damage to sperm and egg cells
10. death

Not all symptoms occur in people exposed to radiation, nor are the responses of two people exposed to the same radiation necessarily the same. The seriousness of the symptoms generally depends on the length and strength of the radiation exposure received. As you might expect, the higher the dosage, the faster these effects occur. Assuming you survive the immediate exposure to a serious radiation event, long-term effects due to mutations (unintended changes) of your genetic material that are not immediately repaired or removed can occur years and even decades later, and in your offspring. Here are some of the delayed responses to severe radiation exposure.

- graying hair
- cataracts
- over twenty different types of cancer (malignant tumors)
- growth of benign tumors
- damage to reproductive organs
- damage to offspring conceived from sex organs exposed to radiation in space

IMPACTS IN SPACE

We have seen that radiation, in the form of photons, atomic particles, and electrons, has the effect of damaging individual atoms or small groups of atoms with each impact. When the impacting body is larger, at least the size of a dust mote, effects occur on more macroscopic scales. If you were asked to choose one word that summarizes the history of the solar system, it should be "collisions." From the moment the solar system began coalescing out of a small piece of an interstellar cloud of gas and dust, innumerable collisions have occurred. Atoms, molecules, and then pieces of dust in that cloud started running into and adhering to each other. Bigger and bigger bodies collided. Those objects moving fast enough shattered each other, while slower collisions caused bodies to stick together. Over a hundred million years or so, the cloud of gas and dust coalesced into a few large bodies, creating the Sun, planets, moons, and larger asteroids. Billions upon billion of smaller pieces of leftover debris remain in the solar system. These are potential villains in the realm of space travel.

Fast-forward to today. You may be surprised to learn that debris from space is continually entering Earth's atmosphere. However, we are safe from daily impacts here on the ground because the atmosphere heats and vaporizes the frequent, small (dust- and pebble-sized) space debris. Some mass is also vaporized off larger pieces of debris falling through the atmosphere, which can survive to land on the ground and cause damage or injury in the process. Space debris on Earth are called meteorites. Stories abound of people, animals, and buildings being struck by meteorites. While some of them have been documented, many are anecdotal.

Impacts in Low Earth Orbit

Impacts by space debris, both natural and human made, have been observed on spacecraft in low Earth orbit. The atmosphere's protection is gone by the time we reach the altitude of the *International Space Station*. The Van Allen belts deflect some of the lowest-energy particles from space, provided that these particles are electrically charged (typically by having lost electrons). Magnetic fields don't deflect any electrically neutral particles flying through them, and the Van Allen belts are not strong enough to significantly deflect more massive pieces of space debris, pebble-sized or larger. Therefore, debris from space strikes any and all human-made equipment in orbit around the Earth. This is made vividly clear from the impact craters seen on pieces of spacecraft and space stations that have landed on or fallen back to Earth. After less than a decade's exposure to impacts in space, some of them, initially smooth, are cratered like the surface of the Moon (figure 7.2).

Furthermore, we humans create space debris, some of which also strikes spacecraft in low Earth orbit. Such debris includes particles formed

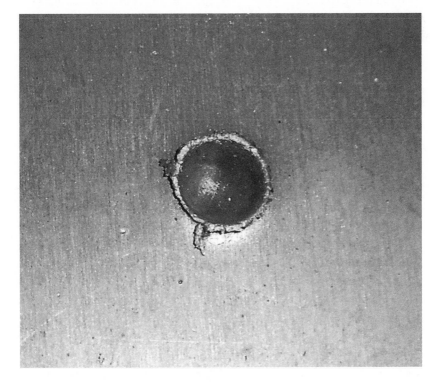

FIGURE 7.2

A cratered piece of the NASA Long Duration Exposure Facility, which orbited the Earth from 1984 to 1990.

NASA Langley Research Center

while solid rocket booster engines are firing, human waste, paint chips flecked off spacecraft, glass from solar cells that have been cracked by the impacts of meteoroids or by other human space debris, and from many other sources.

Fortunately, for your enhanced safety, we have the radar technology to track objects in low Earth orbit that are larger than about 0.25 cm (0.1 in) in diameter. When pieces of debris are heading toward maneuverable spacecraft, the spacecraft are moved to avoid collisions. Although collisions with these larger pieces are extremely rare, there are estimated to be more than a hundred million pieces of debris in orbit that are so tiny we can't yet track them and that, because of their larger numbers, are more likely to hit spacecraft. An estimated 500,000 pieces of debris about 1 cm (½ in) or larger exist in orbit, adding up to 4 million pounds (2,000 tons) of material. Impacts from debris of all sorts occur all over spacecraft and space stations. Despite maneuvering the orbiter component of the space shuttle to avoid known impactors, one in eight of each shuttle's windows had to be replaced after each flight because of damage by impacts from untracked debris.

Damage can be caused even by tiny pieces of space debris. While the vast majority of the natural and human-made debris that strike spacecraft in low Earth orbit are less than 1 millimeter (mm) (1/25 in) across, they are moving at such high speeds, between 17,700 and 250,000 km/h (11,000 and 155,000 mph), that they are able to punch holes into spacecraft and other equipment up there (figure 7.3). Meteoroids typically strike spacecraft with a relative speed of 69,000 kmh (43,000 mph), while human-made space debris typically impacts with half that speed. It is so important to understand this danger that impacts at these speeds are now created in research laboratories on Earth to give us a better understanding of what happens when they strike spacecraft.

For a space traveler, it is helpful to understand the holes and other damage to your vehicle or space suit, so you will know how best to respond to these events. There are two kinds of craters created by impacts of micrometeorites that strike objects in low Earth orbit. The first are circular. These are created when the energy of the impact is great enough to cause the surface that is struck to vaporize and explode equally in all directions. In other words, if the impact is powerful enough, a circular

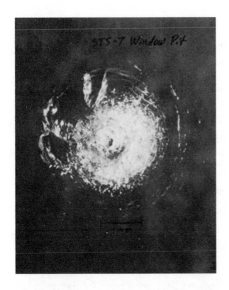

FIGURE 7.3

a) Space debris impact on a window of the space shuttle *Challenger* in 1983.

NASA

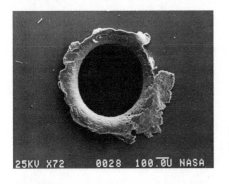

(b) Hole in the *Solar Max* Mission made by orbiting debris.

NASA

(c) A piece of aluminum oxide from a solid rocket motor.

NASA

crater forms no matter the angle at which the incoming body strikes the surface. You can see examples of such craters in figures 7.2 and 7.3b. These events are particularly dangerous because they can cause punctures and enable air to escape from whatever has been damaged.

The second kind of crater is elongated. These are created from lower-speed impacts in which the incoming body strikes the target at an angle (rather than straight down) and material from it and the target primarily sprays forward. This is analogous to what happens when you throw a rock at an angle into water. The incoming object is not necessarily vaporized and can bury itself in the target or even bounce off it. However, even these impacts can penetrate thin surfaces, although they do so less frequently than the more powerful events described above.

Because the consequences of even a pebble-sized meteoroid striking a spacecraft or space suit are so catastrophic, scientists extensively study spacecraft that are visited by astronauts or that return to the Earth to see how they were affected by impacts. One spacecraft specifically designed to collect data on impacts was in orbit from 1984 to 1990 was called the *Long Duration Exposure Facility*. Its surfaces have been studied with scanning electron microscopes. Millions of impact craters have been identified. Another satellite designed to study impacts flew for nearly a year in 1992 and 1993. Over 1,000 impact craters were seen on its 140 square meters (1,500 sq ft) of exposed surface. The largest crater was $2/3$ cm ($1/4$ in) in diameter, while the smallest was about 0.1 cm (0.04 in) across. Luckily, both satellites remained fully operational throughout their missions.

The danger of losing atmosphere into space due to penetration by meteoroids and space debris is so significant that the *International Space Station* has some 200 strategically located protective devices called Whipple shields. They are made of several layers of material, each separated by fractions of an inch to a few inches. This is how they work: the incoming body strikes the shield's outside layer and is shattered into many smaller pieces (along with some of that outer layer). The smaller debris then goes through a series of tough layers made, for example, of Kevlar, the material used in many bulletproof vests. By the time this debris has penetrated the intermediate shields, it has become so much smaller and lost so much energy that when it hits the layer designed to be protected, the debris bounces off harmlessly. Needless to say, the shields have to be replaced from time to time. Finally, all spacecraft carrying humans also have sensors to detect the loss of atmosphere, as can occur when the craft is penetrated, and all astronauts have been specially trained how to respond to such emergencies.

Impacts on the Moon

Nearly all of the impact craters found on other bodies in space that you may visit are circular.[5] These bodies include Mars, the various moons, and the asteroids. Like the circular craters on objects in low Earth orbit, the craters on natural bodies throughout the solar system form as a result of impacts powerful enough to cause the surface and the impacting body to explode. Fortunately, the vast majority of the craters throughout the solar system that we see from Earth or that you will see from your spacecraft were created over three billion years ago. But not all of them!

Modern impacts on the Moon have been detected in several ways. One method, applied in the 1960s and '70s, uses the fact that when an extremely fast impact occurs, it causes the ground to shake. Such shaking also happens if the ground shifts because of internal motion, as happens here on Earth during earthquakes. Geologists detect earthquakes with very sensitive vibration sensors called seismometers.

In order to study whether the Moon has equivalent moonquakes and impacts, seismometers were placed on its surface by astronauts on the *Apollo* 11, 12, 14, 15, and 16 missions. These seismometers did detect motion on the Moon's surface. By measuring when each detector sensed each set of vibrations, geologists were able to determine roughly where the vibration originated. Some came from inside the Moon, while the rest were caused by meteoroids slamming into the Moon's surface. The seismometers typically detected about 170 impacts per year caused by meteoroids that, had they been weighed on Earth, would have ranged from a few grams to 5,000 kg (about 0.22 to 11,000 pounds). That average measured by the *Apollo* seismometers is far below the total number of meteoroids that actually strike the Moon annually. Most of the debris that hits the Moon's surface is much less massive than those early seismometers could detect. In September 1977, the active lunar seismometers were turned off to help NASA save money.

A second way impacts on the Moon are detected is by the light they give off. People have indeed claimed to have seen flashes of light on the Moon's surface from at least as early as the twelfth century. But it wasn't until 1999 (see figure 7.4) that systematic, simultaneous observations by

FIGURE 7.4

Impacts on the Moon. The labeled dots on this photograph show the locations of impacts that occurred in November 1999 during the Leonid meteor shower. The impacting bodies hit the Moon at about 260,00 km/h (160,000 mph). These objects would weigh between 1 and 9 kg (2 and 22 pounds) on Earth.

NASA

observers in different locations around the world confirmed individual lunar impacts from their light. Since then, observers have made simultaneous observations of impact flashes a number of times.

Verifying the impacts on the Moon today is a challenge. Because the flashes are so short-lived and because telescope observing time is often difficult to obtain, simultaneous observations of the Moon by different groups looking for impact flashes are rare. So how would they know when to look with a high likelihood of success? The answer comes from those beautiful streamers in the heavens: comets. As discussed in chapter 1, each time a comet orbits close to the Sun, some of its ices evaporate, carrying away some of the rocky material that had been mixed with the ice when the comet nucleus formed. While the gases drift out of the solar system, the debris bigger than the size of a small pebble remains in the same orbit as the comet nucleus. After perhaps a hundred close passes to the Sun, all that remains of a comet is this rocky debris, which eventually spreads out over its entire orbit.

Whenever the Earth and Moon cross a comet's path, they plow into some of the solid debris that it has left orbiting there. Hundreds or more pieces of this rock are pulled by gravity onto the Earth and Moon. Here on Earth such events create meteor showers. A meteor is a piece of space debris being vaporized by air friction as the debris penetrates our atmosphere. We see the trail of dust it leaves. Meteor showers are occasions when large numbers of meteors occur, all originating in the same area of the sky (namely, from the direction of the comet debris toward which the Earth is heading). Therefore, the appearances of meteor showers, if not the numbers of meteors per hour that result, are predictable. Some of that comet's debris is striking the Moon. Using this information, astronomers who observe the Moon's surface during periods of meteor showers can also detect impacts, as shown in figure 7.4.

A third way that impacts on the Moon are detected is from the gases they give off. The energy transferred between the incoming body and the Moon is so great that most incoming bodies are vaporized, along with some of the Moon's surface. These gases either fall back onto the surface or drift into space, but while they are above the surface, they contribute to the very thin lunar atmosphere. Amazingly, we have the technology to detect the Moon's atmosphere from orbiting spacecraft. The amount of sodium vapor, one of the major components of the incredibly thin atmosphere there, changes when the Moon is struck by debris from meteor showers.

Meteors are seen in our sky, albeit much less frequently than during a meteor shower, every night of the year, even when we are not near the orbits of dead comets. These meteors are caused by random debris entering the atmosphere. Based on these observations, you should expect to encounter impacts of low-mass objects when you are visiting the Moon. However, it is less safe in general to visit space when the Earth is passing through comet debris than when it is not.

By the time human habitats are set up on the Moon, the radar technology to detect meteors is likely to be good enough to provide information necessary to forecast impacts of the more dangerous debris. When you get up in the morning on the Moon and check the Lunanet for the day's weather forecast, you will first want to know the radiation level from the Sun (analogous to how warm it will be on Earth), but you will also want to know how frequent impacts will be that day. Both parameters will have a bearing on what you can safely do.

Impacts in Interplanetary Space

During most of each journey away from the Earth-Moon system, that is, into interplanetary space, you will be protected from impacts only by the materials incorporated into your spaceship by its designers. Your spaceship will primarily be exposed to impacts from small space debris—meteoroids. Billions of these are in random orbits around the Sun.

Humans tend to leave debris wherever they go. Travelers before you will probably have ejected some refuse from their spacecraft. The good news is that every path to every object outside the Earth-Moon system is different for every single spacecraft that goes out there. As a result, none of the human-made space debris that was dumped there before you came by will be in the path of your spacecraft. This is unlike the problem of space debris in Earth orbit frequently hitting spacecraft orbiting the Earth today.

Impacts on Mars

High-speed impacts occur much more frequently on every square mile of Mars than on Earth. This is because Mars's atmosphere is less than 1 percent as dense as the air we breathe. This thinner air cannot vaporize incoming debris as effectively as does the Earth's atmosphere. Fortunately, the majority of the impacts you will face on Mars are from the dust particles carried by the winds that sometimes blanket the entire planet, rather than incoming space debris. The dust there is composed of much smaller particles than grains of sand here on Earth.

Despite the low wind speeds of 35 km/h (22 mph) regularly found on Mars, large amounts of dust will accumulate on anything exposed on the planet's surface. The impacts that occur are typically too weak to create

SCIENCE AND SCIENCE FICTION

The air is held around Earth by our planet's gravitational attraction on the atoms and molecules in the atmosphere. This total mass of gas presses on everything on Earth's surface with an average pressure of 100 kPa (14.7 pounds/square inch). However, our bodies evolved so that we normally don't feel air pressure when sitting in a room with the windows closed. We do feel air pressure when the wind is blowing on us. Obviously, the faster the wind speed, the harder it pushes us.

Surrounding Mars there is much less air than in our atmosphere. Therefore, the normal air pressure is only about .06 kPa (0.09 pounds/square inch), 160 times lower than what we normally feel here. This also means than when the winds blow on Mars, the pressure we would feel from them would be that much less than we feel from winds on Earth today. The air pressure on the surface of Mars is 0.6 percent as high as it is on Earth. Therefore, a 160 km/h (100 mph) wind there would feel like less than a 16 km/h (10 mph) wind here. Consequently, a major storm blowing on Mars would could not blow a person over, as shown in The Martian.

craters in your clothes, for example, but the wind and dust flow creates an electrical charge that forces the dust to cling to many types of plastics and other manufactured products. The speed of the particles when they strike determines how thick a layer of dust will accumulate. Interestingly, studies show that slowly moving debris creates a layer that continually thickens the longer you are exposed to it, while higher-speed dust, moving at, say, 240 km/h (150 mph), only builds up a thin layer. In the low-speed case, the dust clings electrostatically, like the sheets of fabric softener we sometimes wear without realizing they are there after drying clothes (static cling). The electric charge carried by the dust on Mars causes it to cling in large quantities to many things it strikes. In the high-speed case, the energy associated with the impacts cause as much dust to rub off as to cling on, and so the layer remains relatively thin.

Understanding the effects of impacts on Mars is important for people on the Red Planet. The consequences of the low-energy impacts of winds have been explored on Earth using wind tunnels with combinations of dust and heavier grains that are believed to simulate the debris blowing around on Mars. The results show that the material that clings to space suits and visors is not easily removed. It is a housekeeping nightmare. Brushing it causes it to dig into fabrics and to leave scratch marks on plastic visors. The more the fabrics accumulate dust, the more they wear away with the rubbing of everyday use. The more the visors are scratched, the harder it is to see out of them. Also, dust that covers moving mechanical surfaces creates friction that erodes moving parts. Likewise, dust covering solar panels that create electricity decreases their efficiency.

Cosmic clothing designers alert: impacts and radiation on Mars will cause the curling or wrinkling of fabrics! While this won't be a fashion emergency, it is a problem because repeated impacts cause the exposed surface of the material to tighten up, making it stiffer and more brittle. This not only would be uncomfortable but also could cause the fabric to crack under duress. Likewise, the intense ultraviolet radiation from the Sun, which is mostly blocked by the Earth's atmosphere, passes through Mars's thin atmosphere and continually pummels the planet's surface during daylight hours. Like materials struck by ultraviolet radiation on Earth, materials of all kinds exposed to Mars's more intense ultraviolet "age" as a result, becoming brittle, fading in color, and cracking. Even designer space suits for Mars and other bodies will have to be made of materials that can withstand the blistering attack of that radiation.

Impacts on Asteroids

The damage and injury that can be caused by impacts while visiting asteroids are virtually identical to those you would encounter on the Moon. That is, without atmospheres or magnetic fields to protect them, asteroids are struck by the same tiny meteorites that pelt the Moon. However, because the asteroids possess less mass than the Moon, these smaller bodies are unable to pull space debris toward them as strongly as does the Moon. Therefore, they have slightly lower *rates* of impact, and the incoming debris is not sped up as much as the debris falling toward the Moon.

Impacts on Comets

If you visit a comet, the impact dangers come from several sources. The first risk stems from the debris that was recently released from the comet's nucleus. While the density of such particles is extremely low (compared to the density of particles in a sandstorm on Earth, for example), it only takes one hit to ruin your whole day. One way to lessen the danger is to approach the comet relatively slowly and from the Sunward side, where the impact speed of particles in the coma is low.

Some comets have jets of gas shooting out from them. In regions containing solar-heated gas, these jets can escape where the rocky debris is thinnest. While the density of gas and particles released is again very low compared to the Earth's atmosphere (figure 7.5), they nevertheless pose serious danger because even a few impacts can threaten life. It is therefore preferable to visit comets that do not have jets.

FIGURE 7.5

Jets of gas leaving the nucleus of comet Hartley 2. The Sun is on the right of this image.

NASA/JPL-Caltech/UMD

8

GETTING ALONG IN SPACE

—

Psychological and Sociological Aspects of Space Travel

Imagine that you just had a serious fight with a close friend. What do you do? You take refuge in a quiet place in your home or at a cafe to cool down, think things over, and figure out a way to resolve the problem. You could also talk it over with family and other friends, or do some strenuous exercise, which often helps the brain kick into high gear and resolve problems. If the issues can be reconciled, you could then talk to the person and work out your differences. If not, you might choose to "unfriend" the person, avoid them assiduously, and rebuild your life without him or her.

However, what if you are living in an environment where you are never more than 30 m (100 ft) from that person and you have to see him or her several times every day? Imagine that your quarters, perhaps the size of a telephone booth, are only a few meters or yards from that other person's quarters. In that situation, as would occur on a spacecraft or space station, even to think clearly about such a wrenching event you have to wear sound-blocking headphones to mute the background noise of people talking, air conditioners humming, machines whirring, intercoms blaring, and other equipment carrying out their normal functions that otherwise keep the room uncomfortably loud. Worst of all, you know that your family and closest friends are literally a million miles away. At distances beyond the Earth-Moon system, the speed of light will limit conversations you have with people back home, creating a potentially uncomfortable lag of tens of seconds or more between when you say something to them and when you hear their reply.

Let's consider a variety of the personal and interpersonal problems that you might experience on your trip in space. This is not borrowing trouble but an effort to work out plausible issues before they become potentially dangerous. Some of these are events and situations that will affect your mental health. Some are events and situations caused by your mental health. The behaviors and norms described here are derived from studies (by others) of hundreds of ensembles[1] of people in continuous close contact in similarly isolated environments such as wintering over on Antarctica, in submarines, and in voluntary, small-group isolation experiments such as the Mars Society's Mars Desert Research Stations. It is likely that many of the group dynamics presented here—the interactions between pairs and larger numbers of people in self-contained ensembles—will occur on your trip, but *specifically* which ones will occur is virtually impossible to predict beforehand. However, it is good to be prepared.

Because you have *chosen* to travel in space, I have decided to exclude research on people who are forced into restricted environments, such as prisoners. While there are some parallels between the restrictions imposed on you in space and on criminals in the penal systems of the world, the differences are far greater. Because of this, I believe that such comparisons would be misleading.

The complexity of human interactions suggests that any two different ensembles with the same number of people, given the same long mission and the same equipment to carry it out, are likely to have very different group dynamics during the times they are together. The larger the number of people involved, the more likely that the trip will, overall, be a positive experience.

Real-time mental health care support is necessary for the success of a long-term space mission. Because of the profound complexity of human interactions, it is essential that your spacecraft have on board at least one person who is highly trained in working with individuals, couples, and teams on personal and group issues. Ideally, your ship's crew will include a psychologist, or your ship's physician will be well trained in diagnosing mental illnesses, counseling, and dispensing medications that could help alleviate their symptoms of mental illness. It would also be valuable to have a psychologist or other trained mediator who can spot potential personal and interpersonal problems as they are developing, work to defuse them, teach relaxation techniques, and provide counseling to help people resolve these problems.[2] Your ship's captain must have competence in working with people. She or he will have several vital roles, of which ensuring that people get along is one. Finally, the crew will need people trained to provide security in the event that personal or interpersonal conflicts cause someone to hurt themselves or others. Of course, the people who will police the group have the potential to develop any of the same problems, which complicates matters.

THE IMPORTANCE OF SCREENING

Having the privilege of flying in space as a tourist in this first century of commercial space travel will require more than money and good "connections." The farther you want to go away from the Earth, the more you will have to be tested and trained. Trips of a few days to a space station or a week or two on the Moon require relatively little acclimation to living in tight quarters with small groups, whereas trips beyond the Earth-Moon system will require considerable adjustment.

It is likely that people planning to visit the moons of Mars or to hop a ride on a comet will have to spend many months undergoing physical and mental testing and training prior to leaving, as introduced in chapter 3. It is essential that every person going into space be able to endure the physical rigors of space travel as well as the emotional experiences that result from being in close contact with strangers and separated from family, traditional friends, and the variety of activities available here on Earth. Despite all the other requirements of space travel, the need to be able to coexist with fellow space travelers is paramount. Otherwise, the trip could deteriorate into chaos.

A variety of screening processes for people going on long missions in small groups have been developed by navies of the world, space agencies, companies that send people to isolated work sites such as oil rigs, and countries that maintain presences in the far north and on the continent of Antarctica. Screening involves a variety of physical and psychological tests, some quite stressful. It is also possible that by the time you want to travel in space, the space development community will also be using genetic screening. Such tests can check for markers in your DNA that indicate the likelihood that you have or will develop certain physical or mental health problems that would exclude you. Such screening is a hurdle that you will have to pass, and it is important that you react positively whether you get through it or not. If you don't "make it," rest assured that the physical and mental health communities have good reasons concerning your safety and the safety of others to keep you on Earth.

If you do pass the screening, that is no guarantee that everything will go well physically, emotionally, and socially for you in space; preflight testing is necessary, but not sufficient, for ensuring a healthy, happy experience. Indeed, several highly tested, highly trained, highly disciplined, highly motivated, and highly educated astronauts and cosmonauts have fared poorly in space. Here is a summary of some of the published emotional problems that have occurred. John Blaha suffered from depression on the *Mir* space station in 1996; cosmonaut Vladimir Vasyutin on the *Soyuz T-14* mission in 1985 developed both physical and psychological symptoms; the two cosmonauts on the *Soyuz-21* mission in 1976, Boris Volynov and Vitaly Zholobov, developed interpersonal problems that required the mission to be cut short; and the *Skylab-4* crew of Gerald Carr, William Pogue, and Edward Gibson in 1973 developed such antagonism against their ground crew as a result of being overworked that they refused to work or even contact the ground for 24 hours. This list is not exhaustive. The lessons learned from the work loads assigned, the interactions between the ground crew and the astronauts, and the behaviors of the travelers continually refine the way that astronauts are chosen, trained, and "handled" while they are in space.

LIKELIHOOD OF SIGNIFICANT MENTAL HEALTH ISSUES DEVELOPING ON YOUR VOYAGE

Based on the experiences of previous groups in long-term isolation, we can get a quantitative estimate of the likelihood that during your trip you will observe or experience mental health issues. Medical signs are the outward appearance of a person, while symptoms are changes from normal that a person feels but that cannot be seen. A rash is a sign, while feeling as though bands of steel are tightening around your chest is a symptom. For simplicity, I will hereafter refer to both as symptoms.

According to the book *Safe Passage: Astronaut Care for Exploration Missions*, published by the National Academy of Sciences, between 3 and 13 percent of the people in enclosed environments for long periods (months to years) develop symptoms of mental illness each year. Let's apply that to a trip to the moons of Mars. Suppose that you are on a three-year trip and that the spaceflight agencies screen passengers to the very best of their abilities. Therefore, we can hope that the percentages of mental health-related events are at the low end of this range. Let's assume your ship carries twelve people. Using an incidence rate of 3 percent occurrence of mental illness per person per year, you can expect that at least one of the people[3] on board will show symptoms of mental illness before you return to Earth.

As a category of illness, mental health problems are more common in the general population than any specific physical illness. (This isn't quite fair, of course, because the physical illnesses are considered individually, such as cancer, flu, or appendicitis, while I have put all the illnesses classified as "mental" in one group. But the point is that mental illness as a class of diseases is extremely common and needs to be accepted and dealt with in the same way as are physical illnesses.) Furthermore, even the best trained, most "together" people may experience emotional adjustment and other mental health issues, and even symptoms of mental illness, in space. Depression is common among astronauts who stay in orbit for months. American astronaut John Blaha's depression on board the *Mir* space station, mentioned earlier, was well publicized. His trip got off to a bad start when the two cosmonauts with whom he trained were unable to go for medical reasons. As a result, he had to start nearly from scratch to build relationships with two cosmonauts with whom he had neither trained nor developed close rapport.

The U.S. Navy, which screens and trains its submariners exquisitely, finds that psychiatric illnesses developed by crew members are the second most common reason that submarine missions must be aborted. Anxiety attacks, discussed below, are the most common psychopathology that occurs.

GROUP DYNAMICS

Knowing that keeping you mentally healthy is key to your journey's success, let's now consider some of the things that affect mental health. Working well with your peers is essential. We go through our lives as members of many groups. We associate with groups at work, at school, in recreation, at home, and within our communities. Groups can have positive or negative effects on any ensemble of people. Healthy, cooperative, supportive, trusting, cohesive groups usually don't just happen; people have to work at developing these traits. In 1996 several groups tried to climb Mount Everest. As described in Jon Krakauer's *Into Thin Air*, among other books, a storm descended on two groups, leading to eight deaths, including the deaths of the two group leaders. One of the major problems was that they were "groups" in name alone. They hadn't spent the time necessary to learn how to develop sufficient cohesion, mutual understanding, and trust; how to react effectively to emergency situations; or how to develop the collaborative decision-making responsibilities that enable groups to work. Therefore, they were unable to function well together when a series of problems occurred.

Just as you will have to go through preflight physical and mental screening, you will probably also have to spend preflight time with the others planning to go in order to determine whether any glaring personal incompatibilities exist. This time together would include extensive introduction to the equipment you will use and the spacecraft that you will fly in, learning emergency procedures, and plausibly some weeks living and working as a group in a mock-up of your spacecraft's living quarters. As with other testing, this cannot guarantee that the flight time will be satisfying as a result of positive group dynamics, but it will help eliminate the most conspicuous mismatches.

Groups provide people with a sense of security, belonging, and importance. They tend to be semipermeable. It is often hard to be accepted into a group that has already developed structure and cohesion. Some formed groups even have derogatory names such as "green," "wet," "raw," or "newbie" for new members. Membership in one group can also make it difficult to be accepted in another.

Research shows that negative dynamics in isolated groups, such as those on Antarctica, have a variety of consequences. The separation of the ensemble into essentially disconnected cliques can lead to great tension between them. This, in turn, can lead to excessive competitiveness, bullying, or violence. Isolation from a main group can lead to expression of anger and frustration by the excluded member or members. Equally, isolation can cause excluded members of the ensemble to withdraw, become less effective at their jobs, or develop clinical depression or other mental health problems and further isolation. When things go wrong in isolated environments, group dynamics often force the excluded people into the role of scapegoat, blaming them for problems that they neither caused nor could prevent.

Living in very close proximity for weeks or longer creates tensions for many people. Some of the issues raised during these times are related to our environmental needs, while others are related to personal and cultural differences that will exist in your ensemble in space.

Crowding

The concept of crowding is complicated by two important factors: it means different things in different situations, and people in different cultures have developed different comfort levels with the number and concentration of people around them. The first aspect of crowding involves the amount of space each person requires immediately around him- or herself in interactions with other people. This is called "personal space" or "personal zone." Most people feel uncomfortable when their personal space is being invaded or violated. Depending on the country and society you are in, this space can range from a few centimeters or inches to an arm's reach from your body. If your personal space is on the higher end of this scale, you may have to learn to adapt to less separation in space.

The problem of personal space is made worse by the lack of gravity. While you are floating, your sense of up and down is going to be greatly challenged, as discussed in chapter 4. As a consequence of their

freely floating, you will be seeing and talking to people who appear upside down or sideways. These perceptions change the concept of personal
space. Studies on Earth and the experience of some astronauts show that
when interacting with people who don't appear upright to you, you are
likely to need more personal space. In other words, to feel comfortable,
you will have to be farther from those people whose eyes are below their
mouths than you need to be when seeing the same people from a normal
perspective.

Making matters even more challenging, on board your spacecraft, pas-
sageways will be narrow, sometimes barely allowing two people to squeeze
past each other (figure 8.1). Most other spaces in the ship and on other
worlds will also be cramped compared to the space in a typical house or
apartment. Close proximity also brings sensory information, especially
smells, that can be undesirable. Under such circumstances you may find
someone's odor distasteful, causing you to avoid them. Conversely, you
may find someone's smell or another detail about them attractive when
you are particularly close to them, leading you to want to develop an intimate relationship with that person. If that is not possible, you may find
the encounter, positive as it was, frustrating.

Even in a room with relatively large separation between people, personal spaces can accidentally be invaded as people float around in space.
For example, suppose that while traveling to Mars you are reading an e-
book in a common room. Engrossed in the story, you float freely, brush
against a wall, and thereby begin to rotate without even knowing you are
doing so. Your body pivots, slowly moving your feet into someone's back.
This inadvertent invasion of private space may have the effect of raising
stress, tension, anxiety, and hostility in the room, as well as heightening
the awareness of the sexuality of your fellow travelers.

One interesting caveat about the effects of crowding in Western culture
has to do with the gender of the ensemble crowded together. If everyone
is male, the negative effects just cited tend to increase faster than if the
ensemble is either all female or mixed gender. For peoples in other cultures, the gender dynamics are sometimes quite different: women present
in such groups often increase the tension.

The reality of the close quarters in your ship and elsewhere in space
will require you to often adjust the level of comfort you feel with people

ISS002E6537 2001/06/01 16:46:22

FIGURE 8.1

Astronaut James S. Voss and a space suit moving through the hatch of the Zvezda
Service Module in the *International Space Station*.

NASA

in or crossing your personal space. This means that you essentially "let go"
of your normal personal space and either ignore or even welcome people
into it. Obviously, it is easy to let in people with whom you are close. Regarding all "others," you can become desensitized to the close proximity
of people; it is worth going through such training before you leave.

The other common usage of the concept of crowding applies to large
numbers of people in one place at one time, even if personal spaces are
respected. Crowds can create sensory overload, generating the feeling that
one's behaviors are significantly restricted and that one is living in an environment with limited resources. Usually, you feel the presence of crowds
more if you feel a lack of control of the situation. The consequences of
living in an environment that is frequently crowded include greater effort
than usual to seek privacy, withdrawal from social interactions, and a
decrease in the willingness to help others. Furthermore, people living in

crowded conditions tend to be more easily provoked and respond more aggressively or violently than they would otherwise.

Before we turn to the related issue of privacy, it may be helpful to know that good design elements in your spacecraft can decrease the sensations of crowding. Such factors include colors, fabrics, lighting, seating arrangements, room layouts, the relative "busyness" of the room's appearance, and even the shapes of walls (straight are better than curved).

What, then, can be done on board a spacecraft to alleviate the feeling of crowding? One thing that helps is being allowed to reorganize your environment. For example, if the common areas of your spacecraft have dividers to separate different areas where people congregate, the ability to rearrange the dividers to create different "spaces" can help. Likewise, being allowed to change lighting and the scenes on the large-screen monitors throughout the ship would also help people relieve the tension related to crowding.

Privacy

What does privacy mean to you? For some, it means conditions that provide them with solitude. For others, it is a matter of controlling with whom they have to interact. At times, some people work better alone than in the company of others. Some people associate privacy with the opportunity to have an intimate relationship. It can also mean having a place where you can keep secrets or express emotions such as grief or anger without endangering your standing with others.

The need for privacy and the means of achieving it are partly conditioned by the society you were raised in. For example, in some cultures privacy means not talking, rather than being alone. Some cultures look on people who want to be alone with suspicion, believing that they might be pursuing deviant behavior or other activities deemed socially unacceptable. Understanding the culture's expectations and the needs of the others traveling with you prior to leaving the Earth will greatly help you adjust to your new, close environment on board the ship.

We often deal with the increase in stress and tension caused by crowds, as discussed above, by afterward going somewhere "private" and decompressing. On board your spacecraft you will have a very limited amount of private space. Using the sizes of the quarters available to astronauts now on the *International Space Station* as guidelines, you can reasonably expect to have an enclosed berth that is essentially the size of a large telephone booth (figure 8.2), as noted earlier.

The need for privacy and withdrawal from social interactions is often associated with the need for people to "be themselves." When we are with people, we have social façades that are hard to maintain for extended periods. The emotional exhaustion created by continuous contact causes us to need time alone. Being with people for long periods can also lead to your becoming sensitized to the change in your behavior when you go from private settings into groups, which in turn can cause you to begin to avoid crowds more and more. This avoidance can create the impression in others that you don't like them, causing them to avoid you, which makes it even harder for you to interact with them on the occasions when you

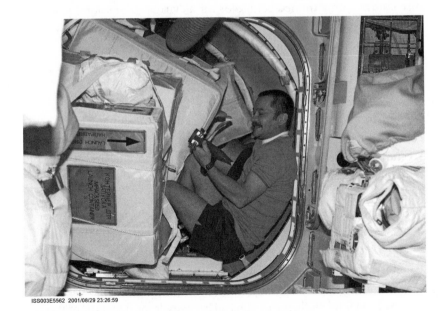

ISS003E5562 2001/08/29 23:26:59

FIGURE 8.2

Cosmonaut Yury V. Usachev in his sleeping compartment on the *International Space Station*.

NASA

wish to do so. As you can see, a whole set of interpersonal dynamics can quickly spiral out of control.

The value of privacy also depends on the length of the confined trip. Experience and studies have shown that for short trips, privacy is a good thing for relieving stress and worry. However, as trips grow to months and longer, too much privacy correlates with increased stress in the group. Even when private quarters are open for conversation, the problem remains. Apparently the group dynamics on long trips require a large amount of social interaction to relieve the stress that normally builds up over time. Too much privacy enables people to "stew" about problems that could be dealt with quickly in a social setting.

People in groups who are experiencing interpersonal problems here on Earth often get together, preferably under the guidance of someone trained in the psychodynamics related to their issue. If talking in groups about problems you are having makes you feel too uncomfortable, private discussions with a trained counselor (or a "virtual counselor") would be good to have. It is not helpful to make a mountain out of a molehill, but it is even worse to let growing emotional and social problems go unresolved.

Territoriality

Related to the sensations of being crowded and needing privacy is the concept of territoriality. We humans are less territorial than many other creatures, but the issue exists nevertheless. In 1938, Admiral Richard E. Byrd (1888–1957), Antarctic explorer, wrote this in his book, *Alone*:

> I knew of bunk mates who quit speaking because each suspected the other of inching his gear into the other's allotted space. . . . In a polar camp, little things like that have the power to drive even disciplined men to the edge of insanity. During my first winter at Little America, I walked for hours with a man who was on the verge of murder or suicide over imaginary persecutions by another man who had been his devoted friend. For there is no escape anywhere. You are hemmed in on every side by your own inadequacies and the crowding measures of your associates. The ones who survive with a measure of happiness are those who can live profoundly off of their intellectual resources.

These scenarios are played out on climbing and caving expeditions, on submarines, and in many situations during wartime.

This need to have and to control a certain space (not just your personal space) has been observed in a variety of studies of groups of people in restricted environs. Think back to the time you were a child, especially if you had siblings. Many children are very territorial with their siblings, resenting any unexpected or uninvited incursions into their room and even more, any changes that others make in it. Fortunately, experience and studies suggest that territoriality is usually confined to spaces that by common consent are the "possession" of individuals, such as their rooms (or berths on the ship) and their office spaces. Many people personalize such spaces with pictures, books, posters, and other personal items. The interpersonal conflict over territory in private spaces often focuses on things at the boundaries.

The territoriality issues associated with common spaces, like a lounge, are different. Groups often try to take over the most desirable common areas, so their actions must be overseen, often by an authority figure such as the ship's captain. Good ship design can minimize this problem by making the boundaries as clear-cut as possible or making them adjustable, so that a group can carve out an adequate, albeit temporary, territory, knowing that the space will eventually be relinquished, readjusted, and reallocated.

Cultural Differences

As mentioned earlier, peoples from different cultures often have different personal and social expectations, needs, beliefs, tastes, religions, and interests. These can lead to interesting discussions and insights (they can be focal points for meetings during the trip), but they can also cause initial discomfort, disagreements, resentment, animosity, and other tensions. Indeed, one of the most potentially explosive problems that could arise in group dynamics is the discovery while in space that someone is a member of an organization that you feel that you absolutely cannot tolerate.

A variety of cultural issues can cause negative feelings, including religions; the treatment of women; the treatment of minorities; attitudes toward privacy; displays of affection; and work and recreation, among others. Since people with a variety of beliefs and attitudes about these things are likely to be traveling with you for months or years in your spaceship, the sooner everyone becomes comfortable with cultural differences, the sooner the group can develop more stable and supportive relationships. (Besides these big issues to be dealt with, many people find that small interpersonal things, like hearing the same story over and over or someone cracking their knuckles, in close quarters can drive them up the wall.)

Some very interesting and surprising results about ensembles have been reported in the literature. In one study, conflicts within ensembles living together for long periods were studied with regard to whether the ensemble was homogeneous or heterogeneous in well-defined ways, the number of people, and the length of time the ensemble was together. The homogeneous/heterogeneous criteria were sex, nationality, age, and experience in similar situations. What they found was that larger ensembles had fewer conflicts and, surprisingly, that conflicts tend to decrease as the mission progresses. Perhaps the most interesting finding of all was that heterogeneous ensembles have lower rates of conflict than homogeneous ensembles.

Why would heterogeneous ensembles together for long periods function more effectively than homogeneous ones? One hypothesis is that homogeneous ensembles start off with the often-false assumption that their members are "alike." Therefore, they think that they don't have to discuss as many things, like politics, religion, and social attitudes, as do people from distinctly different backgrounds. When an initially homogeneous ensemble finally becomes familiar enough with each other's lives, they often discover a variety of differences and incompatibilities. The problem is that these revelations occur well after the point of no return.

If you discover that you are incompatible with members of your ensemble before you depart for space, what then? If, after all the preflight meetings, discussions, and counseling, you can't let go of the negative feelings you have about others in your ensemble, you will be encouraged to go on a different trip.

Leadership

Effective leadership is essential to the success of any space voyage. Your spaceship will have a captain and crew who have trained for years on the intricacies of the ship and of space travel. Several of them will be seasoned space veterans. With a dozen souls on board, your captain has a serious responsibility on her or his hands. Leadership can be handled in a range of styles, from that of a naval ship's captain, who has and uses supreme authority at all times, to an encounter group's facilitator, who makes suggestions and lets the group sentiment move where its momentum takes it.

Based on the experience of long-term isolated ensembles (other than the military, where strict top-down control is mandatory), the role of the captain should not always be authoritarian. In nonmilitary isolated situations, such as wintering in Antarctica, people function better when they are allowed to go about their activities with a minimum of authoritarian direction.

Good leaders, from the ship's captain on down, are sensitive to the dynamics of the ensemble. The leader will learn how to best respond to each member's needs. For example, the leader should be able to identify groups or cliques as they form and to determine whether the new group dynamics are generally positive. Then the leader can determine whether his or her intervention is required to prevent problems from developing in or between groups. Even if your spacecraft is under military authority, the ship's captain has to be an expert in conflict resolution in order to make the flight both enjoyable and safe.

Conversely, leadership requires that the captain slip into an authoritarian role quickly whenever emergencies or other demanding situations occur, such as going into orbit around a destination or docking with another ship. It is as important for you to accept orders without question as it is for the captain to explain as much as possible whenever the circumstances are not dire and time permits.

There is another situation concerning authority figures to which you should be sensitive on your voyage: the effort of some people to usurp power. In isolated groups, one or more people sometimes vie for control, especially during times when other members are withdrawing. People taking authority without permission has soured such groups. This is something that must be avoided in space.

STRESS

You will encounter a variety of stressors during your travel. Stress is your body's response to demands from people and your environment. Change, whether good or bad, is inevitably accompanied by stress. The amount that different situations create varies tremendously, as does the amount that different people can tolerate. Even positive and exciting experiences can be stressful. A little stress sharpens our job performance, pushes us to exceed our expectations, and enables us to compete effectively. As Robert Browning said: "Ah, but a man's reach should exceed his grasp, / Or what's a heaven for?"

The *excess* stress you feel when a situation demands a response that is more challenging than normal is of concern here. Stress comes from three basic realms of our lives: interpersonal relationships, organizational activities, and interactions with the physical world.

Excess stress occurs when we perceive a situation as threatening to or in conflict with our desires, or threatening to our instincts for survival, or beyond our ability to cope with or respond to effectively. When you are in space, it is likely that many stressors will create greater negative responses than the same things would on Earth. Consider the difference between the responses you might feel when you are driving your car, feel it shudder, hear a loud bang, and the motor quits, the lights go out, and you roll to a stop. Now imagine that you are in your spaceship when suddenly you feel a shudder, hear a loud bang, and the ventilation motor quits and the lighting fails.

In extreme cases of stress, our brains generate one of the classic responses: fight, flight, or freeze. High levels of stress can also be symptomatic of specific illnesses, such as depression, anxiety disorders, and post-traumatic stress disorder. Excessive stress can cause many of the following symptoms:

- chronic headaches
- increases in blood pressure and heart rate
- digestive problems
- rashes
- increased anxiety
- increased irritability and hostility
- alienation among group members
- decreased energy
- decreased attention

- decreased motivation to be involved or to solve problems
- decreased productivity
- increased fatigue
- increased need for privacy
- hostility
- boredom
- difficulty sleeping
- impulsive behavior
- obsessive concern about one's own health
- heart attacks

Keep in mind that people who experience excess stress (the kind discussed here) rarely have all these physical or mental symptoms.

Anxiety

We have all experienced anxiety: the feelings of unease, uncertainty, and fear associated with some events or situations. ("Normal" levels of anxiety, along with the physical and mental characteristics that accompany them, have evolved to help us work through challenging situations.) Stress is by no means the only cause for anxiety, which can also be a symptom of other underlying mental illnesses or physical problems, such as depression or drug withdrawal.

The major physical symptoms of extreme anxiety are:

- dry mouth
- shortness of breath
- tightness in the throat
- difficulty swallowing
- sweating
- tachycardia—rapid or pounding heart
- the need to urinate frequently
- abnormal or uncomfortable breathing (dyspnea)
- rapid breathing (hyperventilation)
- nausea or vomiting
- stomach pain/heartburn/reflux

- headaches
- dizziness, feeling lightheaded, and/or fainting
- tremors, twitches, shaking
- fatigue
- diarrhea

Major emotional and psychological symptoms of extreme anxiety are:

- irritability
- insomnia
- anger
- sense of dread or impending disaster
- acute fear of dying
- inability to concentrate
- distractability
- enhanced startle response
- feeling that things are unreal
- feeling that things are out of your control

Post-Traumatic Stress Disorder

Post-traumatic stress disorder (PTSD) has become an epidemic on Earth, in part because of the many soldiers returning with it from conflict zones. It is extremely likely that one or more of the ship's ensemble will experience events on the trip that are so surprising, shocking, disorienting, or upsetting that they develop post-traumatic stress disorder. Such events might include a fight, a death, a life-threatening situation (such as a small meteoroid puncturing the wall of the ship), a mental health breakdown so severe that someone must be confined or otherwise controlled, or news of events back home that is deeply troubling and over which you have no control.

The symptoms of post-traumatic stress disorder typically begin within one to three months of the event and last for at least a month, and often for as long as ten years. These symptoms include:

- recurrent nightmares
- recurrent and intrusive distressing thoughts, images, and memories

- flashbacks to the event causing the disorder
- intense reactions to external or internal cues that resemble the original traumatic event
- reliving the event in your mind
- avoidance of stimuli that resemble or remind you of the trauma
- sleep disturbances, such as insomnia
- hypervigilance
- depression
- irritability
- increased response to being startled
- significant impairment in social and occupational activities
- feeling detached from friends
- diminished interest in activities or avoidance of activities
- sense of foreshortened life
- headaches
- chest pain

Events or stimuli on the trip can also cause PTSD for events that happened prior to your leaving to reoccur. For example, if you witnessed an armed robbery in a flower shop and then smell a certain flower's aroma on the ship that you also smelled during the robbery, you may experience a PTSD episode.

Chronic anxiety or stress can trigger a more serious condition that goes under a variety of names, including chronic asthenia, neurasthenia, Da Costa's syndrome, and soldier's heart, among others. After about two months of chronic anxiety or stress, chronic asthenia causes symptoms including:

- fatigue
- decrease in motivation
- chest pains
- rapid, sometimes irregular heartbeats
- cold, clammy hands and feet
- dizziness
- periodic sighing
- sweating

During chronic asthenia, what was once exciting seems boring and repetitious. Tastes in music and food may change. Members of the ensemble get testy with each other and with the ground crew. Happily, a variety of behavioral and pharmacological ways of treating many symptoms of anxiety exist and should be available to you in space.

CLAUSTROPHOBIA

Claustrophobia looms as a potential problem in the confined environments of space travel. Unless your spacecraft is an extremely high-end model, you will find that the space available to you in it is more limited than any other environment in which you have ever spent more than a few hours. Limited public and personal spaces have the potential to create a variety of mental health problems. A very common one is claustrophobia: the irrational, persistent, and intense fear of confined spaces.

Suffering from claustrophobia leads to anxiety, panic attacks, and impulsive behaviors that could endanger you, others, and the ship. Experiencing claustrophobia while in a space suit would have dire consequences, as you could not remove the helmet in the vacuum of space. A panic attack is a well-defined psychological condition. During an episode, your pulse rate will be much higher than the normal. Normal pulse rates for people ages 21 to 60 are between 60 and 75 beats per minute. If in space you experience a panic attack or any other condition described in this book, or see someone with signs that lead you to believe he or she is experiencing physical or mental distress, you will probably be required to report this immediately to someone who can diagnose the problem and provide help, for the collective good of the ship.

It is useful to know whether you are susceptible to claustrophobia before you leave for space, since you won't be able to step outside (without wearing a confining space suit) or go to a larger room to "stretch your legs" during most of the flight. There are a variety of procedures to test for it. The U.S. Navy puts volunteers into a small pressure chamber, increases the pressure to four times the normal atmospheric pressure so that they feel hemmed in from all sides, and see if the person being tested has a panic attack. Firefighters are sometimes tested for claustrophobia by being blindfolded and put into a small, narrow space and given an assignment, like finding something on the floor. Others are tested by being put into an enclosed suit, like a space suit, with the helmet visor painted black.

As with any phobia, once you experience claustrophobia, you may start to become sensitized to the experience and develop a fear of reexperiencing such confined spaces. If tests reveal that you suffer from claustrophobia or any other phobia, and you desperately want to go on your trip into space, you must overcome the fear. There are a variety of ways to do that with the help of an experienced therapist, including:

- being reexposed to the condition until the fear passes (called flooding)
- learning systematic desensitization techniques
- going into therapy that helps you understand why you feel claustrophobic and teaches you how to turn off the fear (cognitive behavioral therapy)
- watching others in the close quarters who don't have the fear and modeling your behavior on what you see
- taking certain medications that counteract the feeling of anxiety brought on by the phobia

Even if you don't experience full-blown claustrophobia, you may develop symptoms from just being cooped up in your spacecraft for months or years.

TRANCES

If you have been hypnotized, you have been put into a trance. If you haven't, then it is possible that you will put yourself in one during your space voyage. Trances are well-defined hypnotic states. Studies of wintering over in Antarctica suggest that as time goes on, some people there spontaneously go into a trance state, often brought about by the extremely limited environmental stimulation in isolated places. Such changes in mental state are also likely to happen during long trips in space.

During trances, people are able to experience fantasy worlds as though they are real. Both positive and negative hallucinations are common. Such trances enable or force people to tune out the things going on around them. Sometimes people in such hypnotic states appear to be alert and responsive to their surroundings, but their gaze is unfocused and their thoughts are far away. These states become dangerous if people respond to fantasies or hallucinations in ways that can endanger themselves or others, as occasionally happens in the Antarctic, where people hallucinate that they are somewhere else and walk out in the cold to their deaths.

FEAR OF RADIATION

You may well develop fear of the radiation that will be passing through your body on your trip through space. When this response occurs on Earth it is called radiation phobia. Phobias create anxiety related to the object of the phobia before anything has happened, which leads to avoidance of situations in which the feared events can actually occur. However, fear of radiation in space is not a phobia inasmuch as the concern about the radiation to which you will be exposed on your journey is well founded. As discussed in chapter 7, your spacecraft, your space suit, the space stations you visit, and the buildings on a space object's surface cannot protect you from the radiation in space nearly as well as the Earth's atmosphere and magnetic fields do when you are here. For example, during your trip into space you will see "stars" when particles pass through your optic nerve. Traveling through the potentially lethal bath of radiation day after day, knowing that there is little you can do about it, can create a feeling of helplessness that may cause anxiety, depression, or even psychosis.

ATTACKS ON YOUR SENSES

Another set of issues that can lead to psychological and social problems while you are in space relate to the environments you will experience on the ship and in other habitats. These are issues centered on your senses, processing their input, and the resulting changes to your physical and psychological well-being. While I present the environmental issues as a group here, bear in mind that they lead to a variety of responses, including stress, anxiety, boredom, and aggression.

Food

In virtually all isolated environments, whether on a luxury cruise ship, in Antarctica, or on the *International Space Station*, good food (especially chocolate) is one of the things that cheers people up most. However, considering the amount of energy necessary to run refrigeration, it is unlikely that your spaceship will have freezers and refrigerators capable of keeping large amounts of food garden fresh for months or years. Nevertheless, food has improved greatly since the *Gemini* and *Apollo* astronauts squeezed their meals from a tube, ate freeze-dried ice cream, and drank Tang. Now packaged foods are at least hydrated so they have the tastes and textures you expect.

Two complications arise about the choices of foods you make prior to your departure: the amount and the types. Each person in the *International Space Station* is provided with 1.25 kg (2.8 pounds) of food for each day they are in space. The food is precooked and requires no refrigeration, and some of it is dehydrated, just requiring you to add water (just like the stuff you can buy to take camping). The desire to eat in space initially tends to be very limited, as most people struggle with the nausea and malaise created by weightlessness. Therefore, not all of this food is consumed. With the return of appetite, the desire for good, if not great food increases dramatically, only to decline again after months in space.

Planning menus prior to leaving Earth is a double-edged sword. You certainly want to have foods on your trip that you will enjoy, but considering the limited cargo space in your ship, it will not be able to carry the range and quantity of foods that a cruise ship stores, for example. Taking the time to choose foods you like before leaving makes sense from this perspective. However, besides eating less, many astronauts find that their sense of taste changes in space. This occurs, in part, because the fluids in the human body shift location in microgravity, as discussed earlier. In particular, with more fluid in your head, you may feel flulike symptoms continually, giving you a stuffed-up head and a decreased sense of taste.

Traveling in a ship with artificial gravity would prevent this problem, but it is still likely that you will want spicier foods than you are used to on Earth. Fortunately, carrying extra spices is easier than carrying extra food.

Smells

Smells create a wide range of emotions, from lust and hunger to fear and loathing. Our sense of smell, the least-well understood of our major senses, evolved to help us locate nearby desirable things, like food and mates, and to avoid nearby dangers, like fires or toxic chemicals. Smells are also a very effective way to evoke memories that are usually inaccessible to us otherwise.

Historically, closed spaces such as submarines and below decks on ships were notorious for their foul odors. This has changed. Happily, space agencies have gone to considerable lengths to keep odors in spacecraft under control. One common source of smells is an effect called outgassing. When plastics, among other manufactured products, are made, they have molecules that are weakly attached on their surfaces that come off over time. This is what is happening, for example, when you take delivery of a new car and smell the "new car smell." Opening the doors and windows, along with the air outlet for the ventilation system, will eventually diminish these often-pleasant odors. Spacecraft designers work hard to minimize outgassing, but they are not always successful.

People too give off a variety of odors, some attractive, some not. Depending on the hygiene regimen in your spacecraft, as well as rules about wearing fragrances and using scented soaps, some people will have odors that will put others off. This issue will have to be dealt with gently but effectively, so that these people are not ostracized and don't polarize the ensemble. It will also be important in space to avoid diets that produce excessive flatulence.

Unfortunately, people are not going to be the only living things in the habitats of space. Even today, spacecraft harbor mold, bacteria, and other microorganisms. In sufficient quantities, some create musty smells, and some cause allergic reactions and illness. Malfunctioning equipment often creates noxious odors. In 2002 a malfunction in cleaning equipment in the *International Space Station* created a funky smell that required the station's occupants to evacuate part of the facility until the problem was fixed and the air cleaned.

Acting as an early warning system, your sense of smell will also help you discover odors that shouldn't be there, such as gases that escape from equipment leaks. However, in microgravity, many people have a decreased sense of smell, due at least in part to an increase in nasal congestion. The good aspect of this is that you will become less sensitive to offensive odors; the bad aspect is that you will also be less sensitive to good smells, dangerous smells, and tastes. The latter problem occurs because your sense of smell is a major part of your sense of taste.

Temperature

The temperature of our environment is a major factor in determining our level of comfort. Among the sensory inputs that can cause great discomfort are temperatures outside our extremely narrow comfort range. With the possible exception of your sleep quarters, you are unlikely to have control of the temperature of your environment in your spaceship and elsewhere in space. Since different people like different temperatures, it is unlikely that everyone is going to be satisfied. We respond differently to temperatures that are too high or too low. If the air is too warm, you will have the option of wearing lighter clothing. This will help up to a point, of course, beyond which you will begin to perspire, become thirsty, and need to clean your body more often than if you didn't sweat as much. The need to use water for drinking and cleaning is not trivial, as water will be such a valuable commodity in space.

As on Earth, the environment in which you travel in space will determine how the temperature of your habitat will be controlled. Free heating will be provided by the Sun in low Earth orbit, as well as at places on the Moon and at asteroids or comets when they are closer to the Sun than is the Earth. Indeed, spacecraft and habitats in these places will become unbearably hot if they are not shaded from direct sunlight or artificially cooled. Under these circumstances the air conditioning will require the use of energy, which can be expensive if the fuel to generate it has to be carried on board as a nonrenewable resource. Powered by solar panels, the *International Space Station* is very well climate controlled, with temperatures held constant within a degree or two even though that space station cycles between blisteringly hot daylight and blindingly cold night every forty-five minutes.

Closely associated with the temperature of a habitat in space is its humidity (the amount of water vapor in the air compared to the maximum amount that the air can hold at its present temperature and pressure). People are uncomfortable with too much or too little moisture in the air. When the humidity is too low, skin gets rough and dry. When the humidity is too high, it is easier to sweat, and mold and mildew grow rapidly. While there is a range of relative humidities in which people are comfortable, not everyone is comfortable with the same level.

BOREDOM AND MORALE

You're traveling in space. It's amazing! However, the excitement of preparation, departure, microgravity, new group dynamics, high technology, and the exotic beauty of space will begin to ebb as the weeks pass, and with it, your morale will decrease. During this time, you and your companions will begin a complex choreography of interpersonal and group interactions. Some of these will be positive, others not. Sometimes people in such groups get to know everything about each other to the point where the slightest motion or comment will lead to "obvious" conclusions. As Admiral Byrd wrote in *Alone*:

> It doesn't take two men long to find each other out. And, inevitably, this is what they do whether they will it or not, if only because once the simple tasks of the day are finished there is little else to do but take each other's measure. Not deliberately, not maliciously. But the time comes when one has nothing left to reveal to the other; when even his unformed thoughts can be anticipated, his pet ideas become a meaningless drool.

There is no sexism here, as all of Byrd's companions were men. Using other confined groups as examples, you can expect that within months or less, boredom will set in. Have you ever been bored with a movie you are watching? Such situational or reactive boredom is not at issue here.

The boredom that you are very likely to face on a long voyage in space is chronic. Its symptoms include the following:

- a feeling that time is passing very slowly (this is sometimes called subjective time and is discussed in detail directly below this section)
- feeling fatigued and drowsy
- being unable to make plans or strive to achieve goals
- spending a lot of time thinking about yourself
- wishful thinking
- fantasizing and daydreaming
- impulsivity
- excessive sleeping

Boredom can be more dangerous than you might think. The daydreaming, fantasizing, and impulsivity associated with it can become so pervasive that they can lead to actions or behaviors that are unacceptable or even dangerous to the ensemble. For example, you might fantasize about going outside—for just a minute—to see what "space" smells like. Of course, this is impossible and furthermore, spur-of-the-moment excursions are verboten.

Often chronic boredom comes from a lack of stimulation or variety. Resolving the boredom in environments such as your spaceship will center on activities that are varied, interesting, and most importantly, *meaningful*. In other words, suppose you have worked through the 27 computer games that had interested you so far on the trip. You find playing another computer game unsatisfying. However, undertaking a project that will culminate in a meaningful contribution to the group is a surefire technique for banishing boredom. Indeed, theme parties and performance, such as a musical, a drama, or comedy also often break the boredom. These are very popular in Antarctica. Taking a sanctioned space walk, especially if there is something to do outside, such as fixing an antenna damaged by an impact, promises to break the tedium. The only problem is that the experience of floating outside is so alluring to many astronauts today that the mission controllers often have trouble getting them back inside!

Many people who become chronically bored cannot "dig themselves out" of the hole into which they fall. As noted, it is essential to have someone on board trained to look for signs of behaviors that indicate negative mental states or mental illnesses and to work with the people who have them before they create crises that can endanger either themselves or the entire trip.

These issues of boredom and morale are so important that in all likelihood the space agencies that will control civilian travel in space will require *everyone* to work or to take courses while traveling through the solar system. Such work, assigned in moderation and executed carefully, has been important to astronaut and cosmonaut morale to date. There really is no substitute for it if boredom is to be avoided or minimized. It may also go without saying that establishing and keeping a routine is crucial to mental health and stability. Assuming that this is how space travel operates in coming years, I urge you to find a really satisfying "fit" between your areas of competence and interest and the work and education opportunities available to you.[4]

Physical activity is also a valuable tool for fending off the effects of boredom and improving your mood. Your spaceship will have a variety of both aerobic and muscle-strengthening devices that will reinforce both your physical health and your mental health.

Subjective Time

Your normal sense of the passage of time depends on what you are doing. Reading a good book, you can emerge hours later without having noticed any time passing, while driving somewhere new often seems like it takes forever, when it may only be a matter of minutes or hours. Most of us are used to adjusting our perception of the passage of time. However, as trips in space (or on submarines or in other isolated environments) get longer, more complex, and more dangerous, time becomes even more subjective. Seven factors have been identified that correlate with astronauts' changed concept of time:

- increasing distance from possible rescuers (more than at any other time in your life, you are on your own while on trips into deep space)
- heading toward the unknown (extreme cases of going somewhere new for the first time)
- relying on a self-contained environment (your spaceship)

- increasing difficulty communicating with "civilization"
- increasing reliance on the ensemble for all your physical, emotional, and social needs
- increasing independence from technological resources outside the ship
- decreasing resources for life support and enjoyment as supplies diminish during the trip

The passage of time can slow down so much that a minute seems an eternity. You may begin to feel like a prisoner. The more isolated from each other people become on long trips, the worse the problem becomes.

Impulsivity

A number of common psychological problems and mental illnesses cause people to become excessively impulsive, which could theoretically endanger you and others. For our purposes, the most important cause of such spontaneous behavior is boredom, which can lead people to become thrill or novelty seekers. Also, people with adult ADHD (attention-deficit hyperactivity disorder), among many other illnesses, often become impulsive. Symptoms of impulsivity include:

- extreme distractibility
- difficulty getting organized
- difficulty listening to others
- restlessness
- being mischievous and playing pranks on others (some of which could be dangerous to themselves, others, or the ship)

Besides annoying and alienating others, impulsive acts in space can have disastrous results. What if you impulsively press an unmarked button just to see what happens?

Homesickness

Homesickness is an acutely powerful emotion that can waylay even experienced travelers. In fact, virtually everyone will fall victim to it during the trip. While sometimes homesickness occurs at the beginning, it can evolve slowly and insidiously, eventually becoming apparent when people develop either a permanent or a temporary detachment from the ensemble. Here are symptoms of homesickness:

- loneliness and isolation
- sadness and depression without any immediate causes
- anxiety or panic attacks
- feeling overwhelmed or out of control
- irritability and intolerance of others
- anger at and jealousy of the people you have left
- changing sleep patterns
- overeating or loss of appetite
- malaise
- inability to concentrate
- constriction in your throat, stomach, or chest
- feeling that your senses have changed—tastes and smells are different
- feeling nostalgic for the way life was "back home" and preoccupied with thoughts about home, both the environment and the people

Often, time, activities, and new friendships take care of homesickness. However, many people have such strong reactions that they become depressed and withdrawn, begin to "act out," or start abusing alcohol and drugs (if available). However, loneliness and homesickness can be overcome, often with the help of a counselor, even on long missions in space.

Withdrawal and Isolation

Almost everyone likes to be alone from time to time. But when a person virtually shuns association with others or when the frequency or the duration of the isolation episodes changes, others should be concerned. Those who emerge from their berths only for essential activities or who withdraw from social and close emotional ties should raise a red flag. A variety of situations and mental illnesses cause people to withdraw socially and emotionally:

- agoraphobia (the irrational fear of being in crowds)
- alcoholism and drug abuse[5]
- a variety of personality disorders, such as antisocial personality disorder, avoidant personality disorder, paranoid personality disorder
- depression and related disorders
- obsessive-compulsive disorder
- panic attacks
- schizophrenia
- incontinence
- failing senses, such as hearing or sight (often in older folks)

As with all the mental health issues discussed here, someone on the ship should be trained to spot and treat problems such as these to protect and facilitate the success of the journey.

Despondency and Depression

As in other isolated environments, such as Antarctica, we can anticipate that space travelers will undergo periods of depression. We have all felt despondent or depressed from time to time. We receive a bad grade in school, get a bad job evaluation, or have a disagreement with a loved one. These experiences, called situational depression, tend to last a few days or less and pass when the problem is resolved. Such events will happen to you on your journey. The depression that is considered a serious mental illness typically goes much deeper and lasts much longer. This is the long-lived despondency called clinical depression.

A clinically depressed person exhibits a variety of symptoms that are missing from situational depression. According to the U.S. National Institute of Mental Health, nearly 10 percent of American adults suffer from some kind of depressive illness every year. About a quarter of women and nearly 20 percent of men will experience clinical depression sometime in their life. Besides being common in the general population, depression is a component of the lives of many astronauts, cosmonauts, submariners, people wintering in Antarctica, and others in isolated groups.

Depression is thought to be influenced by a variety of factors and events, often occurring in conjunction with one another:

- family genetic history
- physical, emotional, and sexual abuse either as a child or as an adult or a traumatic separation from a parent
- chronic, severe, or prolonged illness, worry, or stress
- anxiety
- poor self-image
- environmental factors
- having had a previous depressive episode
- terminal events, such as grieving due to deaths or the ends of relationships
- persistent, unresolved problems
- physical trauma such as an accident involving the loss of function or actual loss of limbs
- some prescription medicines
- some physical changes and diseases, including heart attacks, cancer, hypothyroidism, and Parkinson's disease, as well as normal hormonal changes for women that may occur during childbirth and menopause

Since depression can stem from so many factors, expert intervention and diagnosis are very important. A variety of mental illnesses are classified under the umbrella name of depression:

- major depression, which disrupts one's ability to eat, sleep, work, study, and generally enjoy life
- dysthymia, a less debilitating depression that nonetheless prevents one from functioning well or enjoying life
- bipolar disorder (formerly called manic depression), which is characterized by extreme mood swings from episodes of depression to episodes of extreme highs, called manias; these cycles can occur slowly, over weeks or months, or over periods of the day (rapid cycling)
- cyclothymia, a mild cycling of mood
- seasonal affective disorder, also called "winter blues," a depression that is usually associated with a decrease in the number of hours of sunlight

You can get a sense of the depth of depression from William Styron, who wrote,

In depression this faith in deliverance, in ultimate restoration, is absent. The pain is unrelenting, and what makes the condition intolerable is the foreknowledge that no remedy will come—not in a day, an hour, a month, or a minute. If there is mild relief, one knows that it is only temporary; more pain will follow. It is hopelessness even more than pain that crushes the soul.

Mental health issues such as depression are made more complex by the fact that some people have exactly the opposite symptoms of those experienced by others with the same illness. For example, some people experiencing depression put on weight, others lose weight; some can't sleep, others oversleep. Furthermore, many people with these various depressions typically will show many symptoms, while others will show few. Here are some of the most common symptoms associated with depression:

- irritability, restlessness (often the earliest signs)
- feelings of sadness, anxiety, emptiness, hopelessness, pessimism, guilt, helplessness, and/or worthlessness
- loss of interest and withdrawal from things that were once enjoyable, such as sex, hobbies, friends, and other activities
- insomnia, oversleeping, early waking
- fatigue, headaches, chronic pain, digestive disorders, and other physical symptoms that don't respond to the usual treatments
- loss or increase in appetite associated with weight loss or weight gain, respectively
- suicidal ideation
- psychotic episode
- changes in hygiene
- changes in behavior

In the list of depressive illnesses above is bipolar disorder, another treatable mood disorder. It is mainly characterized by alternating periods of depression and euphoria or mania. Even with screening, the possibility of someone having this or another mental illnesses on board is high. Common symptoms associated with the manic or "up" side of a bipolar episode include:

- feelings of superiority
- a significant increase in energy and less need for sleep
- increased sexual desire, sometimes to the point of promiscuity
- excessive irritability and aggressiveness
- extremely strong feelings of elation, of feeling a "connectivity" between objects that seem quite separate to others
- racing thoughts and racing conversation
- grandiose ideas, i.e., you've solved a major issue for mankind
- poor judgment and poor social behavior
- drug abuse
- feeling that "God" is directing the manic behaviors
- psychotic behavior and thoughts
- denial that there is a problem; lack of insight into your problematic behavior

Sometimes people suffering from bipolar disorder experience symptoms of both mania and depression at the same time, a situation called a mixed state or mixed mania. Because such people have the "down" feelings of depression and the energy to commit acts that are risky and self-destructive, they are prone to substance abuse and have a high risk of suicide.

Grief

Grief is another emotion for which you should be prepared if you are going on a long space voyage. You may never need to deal with it out there, but then again, you might. The feeling of grief can be overpowering. You have just learned that your child has died. Your spouse is filing for divorce. Your elderly mother just broke her hip. Your best friend fell off a cliff and is now a paraplegic. You, of course, may be a million miles away, not to return home for another six months. Compounding your grief is the knowledge that you cannot be there to help, to offer or receive solace, or to share with your friends and family the sorrow you feel.

Grieving involves physical, emotional, cognitive, and social adjustments to an undesirable change. Typically people go through a series of five stages identified by Elisabeth Kübler-Ross, but not necessarily in the order below:

1. Denial: "This can't be happening. They can't be lost!"
2. Anger/Resentment: "It's not fair. It's their fault they got into that mess."
3. Bargaining: "Please, God. I will change if only you bring them back."
4. Depression: "They are lost. There is nothing I or anyone else can do about it."
5. Acceptance: "I won't forget them, but I must get on with my life."

People often swing back and forth among the first four stages before settling into acceptance. How people grieve is determined culturally and individually. Some people learn to grieve on their own, with a "stiff upper lip." Others are taught to wear their hearts on their sleeves. To illustrate this point, consider the example of cosmonaut Vladimir Dezhurov, whose mother died of cancer while he was aboard the *Mir* space station in 1995, two weeks before the end of his mission. Not only was he unable to be home to mourn with his family, but also Dezhurov's grieving on *Mir* was exacerbated by a well-intended fellow space traveler, Norman Thagard, who consoled Dezhurov as an American would normally do, but which is anathema to Russians. It is important to respect the cultural expectations people have under these conditions, but it is also important to understand that you don't have to go through the grieving process alone.

Third-Quarter Phenomenon

Enough people have been on long, isolated trips to establish that most people eventually undergo a significant drop in their morale. This typically occurs in the third quarter of the experience, hence the name "third-quarter phenomenon," which, oddly, occurs regardless of the length of a trip. Signs include increased stress, discord in the group, and disruptive behavior.

People don't normally sink into depression during this period, but rather go from being excited about the adventure to being rather blasé. Homesickness increases, as well as boredom and sleep problems. This phase typically passes toward the end of the trip, when people start anticipating positive experiences they will have back home. Typically, within six months after returning home, the down period of the third quarter is forgotten.

CONFORMITY TO REGULATIONS

Space is the most hostile, challenging, unforgiving environment that humans have ever visited, although people who explore deep caves or mountain summits may argue otherwise. Survival will require the establishment of guidelines for actions and behaviors, both in your spaceship and elsewhere in space, that everyone must follow rigorously. The practical consequences of disobeying the rules that evolve in space may be catastrophic: injury, death, damage and destruction of spacecraft and habitats.

As we all know, people respond differently to rules and regulations restricting their behavior. The degree to which individuals will conform to regulations in space will likewise vary. Of course, a quasi-military atmosphere on some flights will make the style of command and expected responses very clear. Other flights might tolerate a somewhat looser set of rules and responses.

Because dangers exist on any space mission, the captain of your spaceship will sometimes issue orders that must be obeyed instantly, completely, and without question. While military personnel are used to this mode of operation, civilians are distinctly uncomfortable with it. Because space travel can be a matter of life and death, you will be required to learn and obey a large number of rules designed to keep you and the others with you alive and safe. Nevertheless, one or more people in your ship may develop symptoms of mental stress or illness that will cause them to violate the rules and endanger the ship. Therefore, since at least some of the crew will be trained in dealing with these matters, the ship will have a variety of basic safeguards built in to prevent death and destruction in a worst-case scenario. Some of these innovations include airlocks that can only be used if two people simultaneously activate them; computers that control the ship and the environment that are inaccessible to unauthorized personnel; and sensors that give early warning when damage has occurred or when a person attempts to enter forbidden areas of the ship. In other words, Big Brother will be watching you. Your spaceship will possess innumerable sensors and cameras so that the crew and the ground control can monitor all ship activity. A loss of privacy? Yes—you will have to accommodate yourself to this reality. It will be done for your own safety as well as that of the rest of the travelers.

Part III

MAKING THE MOST OF EXPERIENCES IN SPACE

———

9

EXPERIENCES BY DESTINATION

—

n this chapter we explore many of the experiences that you may have at the different destinations in space discussed in earlier chapters. Before starting with suborbital flights, let's consider one of the most frequent activities that you will do during many parts of your trip: taking pictures.

PHOTOGRAPHY IN SPACE

Wherever you go in space, it is virtually certain that photography will be an important activity for you. I anticipate that there will be three types of cameras for use in space, all of which can be used for either stills or video. The first is tablet computers or equivalent devices. They can be used wherever you can push the button, tell them orally to take a picture, or control them through a selfie stick. Indeed, selfies in space were first taken in the 1960s (figure 9.1). The advantage of taking pictures using a tablet rather than a cell phone while you are outside your spacecraft or habitat is that the touchscreen is large enough for your bulky gloves to accurately press the "buttons" on it.

The second type of camera that you can use in space is analogous to a high-end traditional CCD camera. The larger lenses that these cameras have, compared to those on tablets, give you more flexibility in zooming in and taking high-resolution images.

The third type of cameras are called stereo or, equivalently, 3D (three-dimensional) cameras. Our eyes, each looking at things from slightly different angles, provide our brains with the information necessary for us to

FIGURE 9.1

Astronaut Buzz Aldrin taking a selfie while on *Gemini 12* in 1966. In 2014 he tweeted: "Did you know I took the first space selfie during Gemini 12 mission in 1966? BEST SELFIE EVER."

NASA

perceive the relative distance between us and whatever we see. That is, a person with two healthy eyes sees everything in 3D. A pair of photographs taken simultaneously through two lenses separated by about the distance between our eyes, and viewed through a viewer that sends the left image to our left eye and the right image to our right eye, enables our brain to combine those images and therefore see it in 3D.

Stereo cameras that take two simultaneous pictures were invented in 1833 by Sir Charles Wheatstone (1802–1875). When our brains view the two images, one through each eye (through a stereo viewer, colored glasses, polaroid glasses, or even more modern technology), we see depth—3D. If you have ever used a View-Master® viewer or seen a 3D movie, you understand the tremendous difference between regular (2D) photographs and 3D images. As one who has taken 3D still images since the 1960s, I can attest to the value of this technology.

Cameras of all types face several potential problems in space. First is the wide range of temperatures outside spacecraft and habitats. The highest temperature they will endure is 225°F (107°C; remember, the boiling point of water is 212°F [100°C]), while the coldest is about –300°F (–184°C). Going from inside a spacecraft, space station, or habitat to outside will cause a temperature change of over 350°F (175°C) in a matter of seconds. All parts of a camera or tablet must be able to withstand these rapid changes.

Another problem is that cosmic rays passing through the camera can cause streaks to appear on the image. This occurs when the cosmic ray passes through a line of pixels, the light-sensitive elements in digital cameras, causing the pixels to respond as if normal light had hit them. The result is a streak. Computer software can smooth over most streaks. Another way to compensate for them is to have the camera very quickly take three pictures of each thing you photograph. Even if each picture in a series experienced a cosmic ray, these streaks would likely be through different pixels. Software could then delete the streaks, using the untainted regions of the other images to fill in where the streaks occur.

We now turn to exploring the different space journeys.

SUBORBITAL FLIGHTS

Roller-coaster rides pale in comparison to suborbital spaceflights. The initial acceleration you experience as you rocket into space evokes a variety of emotions. It is a rush to be pushed hard into your seat by the rocket behind you. Don't be surprised if that comes along with a variety of other thoughts and feelings:

OMG, it's really happening!

Is everything going according to plan?

What if . . . ?

I can't wait to get into space!

This last one raises an interesting point. Every stage and every event on your journey into space will be a new and unique experience for you. Therefore, *living in the moment is the most effective way for you to savor and remember everything. This applies to all space travel.* Don't worry about what comes next! Assuming all goes well throughout your trip, the more you internalize each moment, the better you will be able to share the experiences, as well as keep them in your memory throughout your life. For example, in this ascent stage of the trip, let yourself feel the force on you. Try to move. Feel the vibrations. Hear the radio chatter. All your senses (you have more than five) will be working overtime, and it is worth tuning in to and keeping as much of that information as possible.

If you are listening to the radio communication during your ascent or watching the screen with the flight timeline on it, you will know when the rocket "burn" is going to end. From that moment and for the next 3½ to 5 minutes or so, you will be weightless on your parabolic path above the

Karman line. During this very short time in space, you will be allowed to free yourself from the seat and push yourself out of it within a few seconds. Weightless, you can float up into the cabin.

There is one essential thing to do to make the trip especially satisfying. *Go to a window and look at the Earth.* By all reports, seeing the Earth from space is breathtaking and can have a profound effect on you. You will see large parts of entire continents, on which you are likely to make out a number of cities. If you are anywhere near your hometown, you will probably look for it. *Look hard at everything.* It is extremely likely that the space plane in which you are traveling will have many windows so that everyone can look out in many directions. Indeed, the company taking you may have a policy that after one minute of looking out a window, everyone must float to another window that gives a view in another direction. These images will stay with you for a lifetime.

If your trip is at night, you will see the lights from numerous towns and cities in different countries and possibly different continents *all at once.* This will give you a perspective on human life on Earth as interconnected in a way that you can never perceive from Earth. The different cultures, religions, sects, nationalities, and languages lose all significance when you see the Earth as a whole.

Looking out the windows is likely to take up virtually all of your time in weightless flight. There are some other diversions your flight company might plan for a few seconds here and there throughout microgravity. These could include, among other things:

- experiencing the difficulties with proprioception in space, such as having you reach for something
- having different people look at each other from different orientations—seeing a face in front of you with the mouth on top and eyes on the bottom can be quite "interesting"
- having each person briefly spin around either along their vertical axis or around an axis running through their stomach
- trying to catch food pushed toward you in microgravity

Sex is going to be a draw for some space travelers. As long as space suits are worn during suborbital flights, sex in space is going to be limited to kissing, caressing, and fondling. The problem with more intimate contact is that there won't be time enough to disrobe—you will only have about four minutes of microgravity. Even when you make suborbital flights in normal clothes, or less, the likelihood of successful coitus is limited for several reasons. First of all, the high gs that you experience on the way up may affect a man's ability to quickly get an erection. Allowing for an erection to occur (perhaps by taking prelaunch Viagra), connecting in microgravity is extremely challenging. Anytime you push against your partner, you will drift away from each other, since gravity is not "holding you down."

There are basically two options for a quickie in microgravity. First is a variety of hand-holds, straps, and belts mounted strategically in the space plane. Both you and your partner would have to use these to remain close to each other throughout the sexual encounter. The second option is for both of you to float into a large bag, perhaps mesh, so that it doesn't feel claustrophobic, that holds you close together. In either case, your loss of proprioception will also be a challenge because you don't want to over-reach as you extend your hand toward your partner and thereby punch them. As with so many things in life, the best preparation you can do on Earth is practice, practice, practice.

Once sex in suborbital spaceflights is logistically possible (i.e., space suits need not be worn), it is absolutely certain that any number of companies on Earth will start designing products that make sex in space easier and more enjoyable. There is a logistical issue to be aware of: space planes are small, and if you want to have sex in space in relative privacy, you may have to book the entire plane and have a screen erected behind the pilots.

There is also a safety issue about sex in space. Once your space plane has entered the parabolic (unpowered) part of its orbit, the path it follows and the time you will be weightless are set in stone. You may have, for example, 4 minutes and 47 seconds of microgravity, after which you will drift down in the cabin as the wings bite into the atmosphere. If either you or your partner is on the cusp of an orgasm at this time, when it will be essential for you to get back into your seats, you may really resist the need to uncouple. Depending on how quickly your space plane decelerates, this could pose a serious problem—the force acting on the two of you could lead to injuries.

Whether or not you have sex in space, the descent back to Earth will be another unique experience on your spaceflight. Although you probably won't be aware of it, as it happens during your weightless time, your space plane will have risen to the highest point of the flight, called its apogee, and begun descending back toward the atmosphere. As the craft enters the increasingly denser air, air friction against the hull and wings will start slowing the descent. Since you are floating, you won't slow with the ship. As a result, you will start drifting down toward the Earth-facing side of it. At that time you will have been instructed to get back into your seat and strap yourself in, because within minutes you will start decelerating so rapidly that you will again be pressed hard into your seat as the craft penetrates deeper into the Earth's atmosphere.

The descent toward the Earth will be an unpowered glide, just as the space shuttles used to do. The wings of your ship will do the heavy lifting to slow the craft down and then allow the pilot (or computer) to guide it home. There may be some fuel kept in reserve to allow for emergency course corrections caused by unexpected winds or other meteorological issues.

The angle at which you descend into the atmosphere will probably be quite shallow so that the heat generated by air friction can be distributed over a large area of the bottom of the wing surfaces and fuselage. As a result, you will fly for quite a while, descending ever more slowly. Watching out the window as you approach the Earth, the change from seeing large regions of the planet to again seeing the smaller parts of it—individual cities, towns, farms, roads, and vehicles, among other things—can help reinforce the perspective you may have gotten in space of the Earth and everything on it as pieces of a single system.

You will glide to a landing, after which a truck with a towing fork, like those that move airliners back from the gate, will be attached and tow you to the terminal, where you will be reunited with your friends and family. Then your postflight celebrations will begin.

ORBITAL AND MORE DISTANT DESTINATIONS

Launching to orbit requires a much more powerful rocket than do suborbital flights. Indeed, you will begin journeys out to a space station on top of a full-fledged rocket, just as all spacecraft sent to the *International Space Station* and beyond are launched today. The total time for the ride on this rocket to your initial orbit, about twice the altitude of the Karman line, is only eight minutes, as noted earlier. Since many, if not all tourist flights beyond Earth orbit will begin with a visit to an Earth-orbiting space station, this section applies to all trips other than suborbital ones. The exception would be flights that for technical reasons are sent immediately from their initial low Earth orbit directly out of Earth orbit, as was done for the *Apollo* missions to the Moon.

The amount of acceleration you experience going into low Earth orbit will be greater than is experienced anywhere on suborbital flights. It can be "fun" exploring the effects of that serious acceleration on your body, such as trying to raise your arm or leg during the high-*g* lift-off phase. Depending on your carrier, there are likely to be cameras mounted in strategic places outside the rocket that will give you live images of the Earth moving away from you as you rise. You can get a feel for how exciting these views can be by watching such videos on YouTube today. However, seeing your ascent occur from the inside of the spaceship takes the experience to a whole different level.

Ascent to orbit includes several "stages" of rocket burns, after each of which your rocket drops off used hardware. As you go through these transitions, you may "feel" yourself being released from the bonds of the Earth. This feeling can be compounded by the change in direction you will be traveling, more and more horizontally the higher you get, until you are moving essentially parallel with the Earth and fast enough to stay in orbit.

Unlike suborbital voyages, where travelers will be encouraged to release themselves from their seats as soon as possible after they become weightless, you will likely be required to keep your seat belt on when you enter the initial low Earth orbit after launch. Why? In the best of all possible situations, booster rockets will send you spiraling up from there to the destination space station within minutes, leading to docking less than six hours later. If, however, your flight farther out is delayed by technical problems, which does happen, then you may be allowed to unbuckle and float around at the first orbit level to which you are carried while things are being worked on.

In any event, you will definitely experience microgravity when you dock with the space station that is either your destination or a way station on a trip farther from Earth. Assuming that a fully effective, nondrowsy, anti-space sickness med has *not* been developed, you can anticipate a few days of discomfort there, as discussed in chapter 4. Depending on logistics, you may stay in that station before going further away from Earth until you have weathered Space Adaptation Syndrome. If meds are available to eliminate that, the transition to microgravity will obviously be much more enjoyable. Nevertheless, during this time you will still undergo the body adjustments described in chapter 6. After this transition period, the fun begins!

Throughout your flight, views of Earth are likely to be entrancing (figure 9.2). You are much higher up than a suborbital flight ever gets, so your views will be broader and more spectacular. Unlike on the suborbital flight, you will also have the opportunity to see virtually every place on Earth both in daylight and at night. You may want to consider tweeting

FIGURE 9.2A

Night view of Italy and countries along the eastern coasts of the Adriatic and Ionian Seas. Image taken through the *International Space Station*'s cupola (figure 9.2b)

NASA

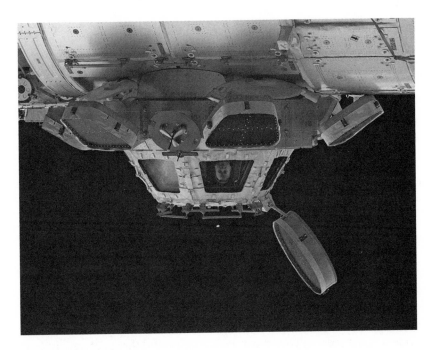

FIGURE 9.2B

View of the *International Space Station*'s cupola for observing Earth.

NASA

as you watch. Many people on Earth will be hungry for the new, real-time insights you can provide about our planet. Conversely, you may want to journal your experiences privately with the potential for polishing them and publishing them or blogging later in your journey or when you get back home.

There will be a variety of interesting and entertaining microgravity-related activities on every trip into space, regardless of destination. Indeed, virtually everything you do can be made different in space than what it is like on Earth.

Eating in Space

We saw in chapter 6 how eating in space is different in many ways from dining on Earth. Every aspect of consuming food, however less-than-truly-appetizing it appears and tastes, can be made entertaining. Here are some examples:

- food can be presented floating right in front of you, which you either pick up or lean toward to eat
- food can be set in containers upside down above your head, so you pull it down rather than pick it up
- friends can gently float food toward you for you to catch in your mouth
- you can put fluids in the specially designed cups described in chapter 6, turn them upside down, and the fluid won't spill out
- you can drink from those specially designed cups
- you can enhance the food flavors by putting onto them much more of the appropriate condiments than you would ever have dreamed of using on Earth

By the way, powdery flavorings such as salt, pepper, sugar, chili, and other spices are supplied in liquid form so you don't inhale them by accident or get them in your eyes if they float around the spacecraft. Indeed, any consumable that would normally leave small pieces, such as regular bread, is reprocessed to avoid leaving debris, or substitutes are found. For example, regular bread would leave crumbs floating around, so it is replaced with tortillas, which are much less likely to do so.

Movement in Microgravity

After Space Adaptation Syndrome passes, adjusting to microgravity can be a lot of fun. In microgravity, you will remain wherever you are when you stop yourself from moving. In other words, if you are in the middle of a room, yards from any surface, you will just hover there. Likewise, if you are floating there and someone starts spinning you, along either the axis running from your legs to your head or any other axis, you will continue to spin until someone stops you or you bump into something. One especially entertaining thing to do is to explore the conservation of angular momentum in microgravity. Angular momentum is a measure of the energy you have as a result of rotating (spinning on an axis through yourself) or revolving (orbiting around something else). Angular momentum is a conserved property of matter. It does not change as long as the spin or revolution of the object is not affected by outside forces, meaning once you have a certain amount of angular momentum by spinning in microgravity, say, it won't change unless you bump into something or someone grabs you.

The consequences of conservation of angular momentum on Earth are best shown by ice skaters. An Olympic skating performance, for example, is often capped with the skaters putting their arms and one leg way out and then spinning themselves around. As they pull these appendages in, conservation of angular momentum causes them to spin faster and faster. You can do the same in microgravity. For example, hold equal weights in your hands, spread your arms way out, and have someone spin you around the axis running down your body. Then pull your arms slowly inward. You will speed up—the more mass each weight you hold contains, the faster you will eventually spin. To slow down, extend your arms back out. To stop, have someone grab your body while they are holding onto something that will anchor them (or you will cause them to start moving around).

Another way to explore the conservation of angular momentum in weightlessness is to put your arms out over your head and extend your legs down as though you were standing with your arms raised up. Then have someone spin you so that your arms and legs are pivoting around your stomach, like a pen spinning on a table. Then tuck in your arms and legs. The closer they get to your axis of rotation, the faster you will spin.

Yet another fun experience in microgravity is to fly across a room like Superwoman or Superman. If you crouch down against a wall and then very gently push your legs against it, you will start drifting across the room. The reason you should not push too fast, until you get used to doing this, is that your body carries with it linear momentum, even in microgravity. Linear momentum is a measure of how much energy you have because you are moving in a straight line. Even in microgravity, your linear motion is associated with how much energy your body has. Therefore, when you reach the opposite wall, if you are moving too fast, your arms won't be able to cushion your collision. You could seriously hurt yourself or any equipment you fly into. You'll need to start slowly and work up to a comfortable, fun speed of flying. Your friends and family will love the opportunity to Skype with you. Imagine them telling their friends, "I talked to my mother (or father) in space! He was floating and spinning around as we talked!" It will go viral!

Hair and Skin Care in Space

Grooming can be a challenge in space. As noted earlier, crumbs and dust are potential problems in microgravity because they can get in your eyes and lungs, as well as into food and sensitive equipment. When hair is cut or shaved, all the clipped hair has to be vacuumed so that it does not interfere with anything on the spacecraft. Likewise, powdery cosmetics are not permitted, although creamy cosmetics can be used. Keep in mind that in microgravity, attached hair floats around, which often makes for interesting images (figure 9.3).

FIGURE 9.3

Astronaut Marsha Ivans having a bad hair day in the *International Space Station*.

NASA

ASTRONOMY FROM SPACE

Even if you are not interested in looking at things astronomical before you leave for space, the opportunity to observe the Moon, the planets, and other objects with the powerful telescope located on a space station, on a spaceship, or at your destination is likely to change that. Observing astronomical objects from space has two distinct advantages over observing them from Earth.

First, since there is no atmosphere up there, it is continuously dark (and there are no nearby clouds), so you can observe objects in space anytime and, depending on your target, they will be visible continuously, except when blocked briefly by the Earth, the Moon, the Sun, Mars, or asteroids. Even looking at areas very close to the Sun from space, with the Sun blocked out, will allow you to see the stars and planets near in angle to it. *Never look at the Sun from Earth or space without an approved filter, since just a few-second glance will begin to permanently damage light-sensitive rods and cones in your eyes, even during an eclipse!*

Second, there is no air in space to distort objects, as the Earth's atmosphere does for space objects observed from our planet's surface. Light from objects in space passes through Earth's atmosphere on its way to the ground. The problem is that the air is in continuous motion, as we know from the wind we feel. This motion is caused by differences in the air density from place to place and height to height. Where the air is less dense, surrounding air pushes its way into that region, and vice versa, creating wind. At the same time, the differences in density cause the light passing through the air to change direction, exactly as a lens does. Since the air changes density as it continually rises and falls and moves from side to side, the direction of the starlight passing through it continually changes direction, an effect we see as twinkling. This is the same effect you see when you look down a road on a hot day and it appears to be shimmering. Twinkling tends to smear out images from space as seen on Earth, making them less distinct. Objects don't twinkle at all as seen from space.

Objects that you should definitely try to see from space are:

- the Earth
- aurorae and meteors in the Earth's atmosphere
- the Moon
- the Sun (through a special filter)
- Mars
- Jupiter
- Saturn
- the Milky Way
- the Andromeda galaxy
- the Magellanic Clouds
- the Orion molecular cloud
- any active comets
- the Pleiades and other open clusters of stars[1]
- a globular cluster of stars

SPACE WALKING

Buying the trip package with the space walk included, while much more expensive, is well worth considering. Walking on any of your destination worlds will be explored shortly; here I specifically mean floating in space outside your space station or spaceship. If you choose this option, you will train underwater for the experience, as described in chapter 4. The decompression preparation on the space station you will undergo is also described in that chapter. You will wear an antinausea patch during your space walk to prevent you from vomiting in your space suit. You can expect to be able to spend several hours outside, tethered to the space station or spaceship. Keep in mind that the expression "space walk" is misleading since most of the time you will be floating in space, not walking. You actually may have the opportunity to literally walk in space. That can be done with magnets in your boots and a surface on the space station or spaceship that attracts magnetic fields. They will have guide wires for you to hold onto, as necessary. There will be several things to try during your space walk, including:

- looking at the Earth—this is most impressive if you are orbiting our planet, but it is always memorable

- if you are in Earth orbit, standing on the station with the Earth directly overhead (in other words, upside down as seen from Earth) so you have to look up to see it
- floating out as far as the tether will allow, in the silence of space
- some simple mechanical activities, such as using a wrench on a bolt, while you are otherwise floating free. It won't work well, if at all, but it will be fun to see what happens.

WORK DURING YOUR VOYAGE

Constructive work is crucial to enjoying extended space voyages. Long trips, out beyond the Earth-Moon system, can become very tedious, as discussed in chapter 8. Even if you choose not to do any work that you normally do—after all, you are on holiday—it might be very helpful both to your state of mind and to others involved in developing space travel if you participate in some research during your trip, for which the company or institution funding the research is likely to pay you. Considering the limited number of people who will have been in space before you, work is likely to be available in virtually every aspect of human endeavor. For example, a pharmaceutical company may be testing a drug that improves the sense of taste in space. The tests of such meds are done in "double blind" experiments. Half of the "double" means that half the people being tested get the med, while the other half get a placebo that has no medical effects on the body. The other half of the "double" means that the people administering the test (giving you the tablet or attaching the patch, as well as collecting information from you) will not know whether you are actually getting the med or not. Consider participating as a contribution to human understanding of Nature.

SEX IN SPACE

Sex is going to become a part of space. When I worked for NASA as a summer fellow at Ames Research Center at Moffett Field, CA, in the mid-1980s, someone high in the NASA hierarchy briefed us about the big picture of space exploration at that time. The speaker then took questions. The first one was, "What about sex in space?"

"Go for it," he answered.

I have not asked any of my astronaut friends about sex in space, and none has volunteered any information on the subject. It has been a "don't ask, don't tell" subject, but in this day and age, I would not be surprised if you start seeing blogs entitled, "I had sex in space" or, "I am now a member of the 400-km-high (250-mile-high) club" shortly after commercial space travel gets under way.

Unfortunately, there is a potential medical problem concerning erection in space, to which NASA may know the answer. Recall from chapter 6 that blood and other bodily fluids are redistributed in a body in microgravity. Since an erection is caused by blood flowing into the penis, does the fluid redistribution in space adversely affect the ability of a man's penis to become erect? If the answer is no, then . . . no problem. If the answer is yes, then supplies of erectile dysfunction meds for men in space will have to be available. Similarly, women secrete to lubricate their vaginas during sexual encounters. This fluid comes from their blood. If it is also decreased in volume by the fluid redistribution that occurs in microgravity, then they will need to have vaginal lubricants available in order to enjoy the encounter.

Allowing that such "technical" problems can be overcome, you will have as much time as you need to have sex in space. The mechanical aids to allow you and your partner to stay together, some of which were introduced in the suborbital section of this chapter, will be there. Another option for keeping you and your partner together is flexible sheets you slip between, which are firmly attached to a mattress fixed to some surface. Indeed, finding better devices for sex in space is a job you might want to consider doing while up there. Welcome to the club.

RETURN TO EARTH

One option for descent back to Earth from an orbiting space station is in a space plane. That journey home will be a longer version, pulling more gs, than the return described earlier for the suborbital flight. It will be a truly memorable roller-coaster ride as you enter the atmosphere, change direction, change speed, and eventually roll to a stop. Another return option is in a space capsule with a parachute. While not as smooth a process as the space plane lander, it will also leave you with stories to tell. We now turn to destination-specific activities to consider while you are in space.

THE MOON

Leaving the Earth's orbit en route to the Moon or any more distant destination may bring back memories from earlier in your life when you were leaving your childhood home for college or a career. Your trip from Earth orbit to the Moon is likely to start a few days after you arrive in space, when the effects of Space Adaptation Syndrome have run their course. Strapped into the spacecraft going to the Moon, you will experience a translunar insertion burn, which is the firing of your ship's rocket to take you out of Earth orbit, quickly through the Van Allen belts, and then on a coasting ride to the Moon. Recall from chapter 2 that this trip will take two to three days.

You will be comfortable on the transit to the Moon. During that part of your trip, you will not have to wear your space suit. From the time your spacecraft's translunar injection burn taking you out of orbit and beyond the Van Allen belts is over until the lunar insertion burn puts your spacecraft in orbit around the Moon, you will experience microgravity. Watching Earth recede and the Moon grow larger will be a major activity on the way to the Moon. Indeed, only when you have left low Earth orbit will you be able to see the entire disk of the Earth (that is to say, all of one half of it). So far, only the *Apollo* astronauts have done that.

Looking at the Earth over a period of hours, you will see different surface features and different cloud patterns as our world rotates. However, looking at the Moon at any time during your trip to it, you will always see the same features—the same side. In other words, as seen from Earth or from space between us and it, the Moon does not *appear* to rotate.

In fact, the Moon does indeed rotate, but it does so at exactly the same rate that it orbits with the Earth,[2] once every 27⅓ days. Because of this synchronous rotation, the same side of the Moon always faces the Earth.

The first time humans ever saw the "far side" of the Moon was on October 7, 1959, when the Soviet *Luna 3* spacecraft flew past the Moon, turned around, and took pictures of it. The far side looks very different than the near side, which we see from Earth (figure 9.4). Whereas the near side has a variety of dark, relatively smooth maria (seas) and heavily cratered, mountainous highlands, the far side is almost entirely covered by highlands.

You have undoubtedly heard the expression "dark side of the Moon." This is often used, incorrectly, to refer to the lunar far side. The dark side of the Moon is the side facing away from the Sun, while the far side is the side facing away from the Earth. So, whenever we see less than a full Moon, we are seeing part of the dark side, but we never see the far side from here. The far side and the dark side are exactly the same only during a full moon as seen from Earth.

Things will get busy as you approach the Moon. You will don your space suit again and strap in before the lunar insertion burn of your ship's rocket engine puts you into orbit around the Moon. Entering lunar orbit is very exciting in large part because you will be going around the Moon and therefore will see the far side, which very few people have ever seen directly. Indeed, through 2016, only 27 people have ever gazed out a window at the lunar far side (see figure 9.4).

Once in orbit, you will either dock to a lunar-orbiting space station from which a lander will bring you to the Moon's surface, or dock directly to a lander. The landing experience will be especially interesting if you can watch it out a window, but it will be glorious even on a large-screen monitor. The Moon's surface, with its gray tones, is covered with a regolith of finely powdered rock (see figure 1.3), dotted with occasional solid rocks. The regolith is sufficiently compact that we can land spacecraft on it, walk on it, and even build habitats on it. The engineering of designing living quarters for the Moon is very much a work in progress.

The Moon is essentially airless, and the force of gravity it exerts is only about a sixth as strong as the force of gravity on Earth. This weak gravitational force makes it easy for present-day rocket technology to bring landers safely to the surface and for orbiters to blast off from it. As your lander descends, its rocket firing downward, a lot of the regolith below it will spray up around you. Descending into this spray will be heart-

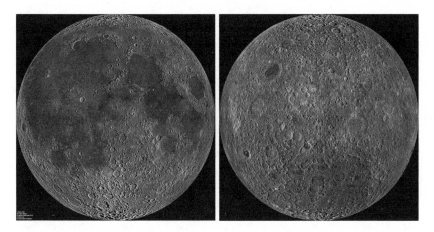

FIGURE 9.4

Near side (left) and far side of the Moon.

NASA

stopping, perhaps looking like a slow motion jump into water as the water splashes to the sides. If all goes well, your lander will settle smoothly onto the surface. Some landers, such as the one that landed on comet 67P/ Churyumov-Gerasimenko in 2014, actually bounced, but such events will become rare as the technology improves.

Because of the dust spray, the landing site will be a considerable distance from your hotel, which is likely to be underground or in a cave in order to help protect inhabitants from incoming space debris and space radiation. If the landing site were close to the habitats, the spray would put a growing layer of debris on the entryways and any vehicles on the surface near them. Putting habitats underground will also make it easier to control their temperatures. Daytime temperatures on the Moon range with latitude from a high of about 250°F (120°C) at the equator to a low of about –350°F (–210°C) near the poles. Nighttime temperatures decrease as the Moon cools, also to about –350°F.

Because of the distance between landing site and habitats, it is likely that you will take a Moon taxi (Moonmobile?) to your hotel. One of the things you will notice very quickly is how dirty everything is there, from the taxi to the space suits to every outside building surface. The reason is that much of the regolith dust is electrically charged; it has static electricity. The regolith dust becomes charged because radiation from the Sun knocks electrons out of orbit in its atoms and molecules. The dust particles are then attracted to virtually everything that we bring to the surface. This dust has the acrid smell of burnt gunpowder, which is quite unpleasant.

Designers may someday develop space suits that shed dust effectively and repeatedly. This will entail more sophistication than just coating everything with Teflon, but the problems will probably be overcome. In the meantime, engineers studying these problems have come up with the plausible solution that you should bring along a large supply of single-use coveralls to cover your space suit and prevent it from aging due to impacts of dust, radiation, and other debris. Likewise, your helmet visor can be coated with single-use peelable plastic layers similar to the single layers of plastic that are used to protect the touch screens of tablets and similar devices.

The virtual absence of an atmosphere adds to the romance of the Moon. The sky there is always pitch black and filled with stars except, of course, in the direction of the Sun or Earth. The stars as seen from airless worlds or from inside spacecraft do not twinkle because twinkling is actually caused by the motion of the Earth's atmosphere, as discussed earlier in this chapter.

During the day, the surface of the Moon will be brightly lit by sunlight, while at night it will be dark. (However, sunlight scattering off the Earth does provide a small amount of light on the dark side of the Moon that is facing the Earth. That is why we can often see the whole side of the Moon facing us even when it is less than a full Moon.)

The day-night cycle on the Moon, which is the same as a cycle of lunar phases (new—waxing crescent—first quarter—waxing gibbous—full—waning gibbous—third quarter—waning crescent—back to new) takes about 29½ days, meaning there are 14¾ Earth days of continuous sunlight on the Moon followed by 14¾ days of continuous darkness.

Solar Eclipses Seen from the Moon

Just as there are solar eclipses seen from Earth when the Moon blocks the Sun, you can also see solar eclipses from the Moon, when the Earth moves between it and the Sun. These events occur on some full moons and correspond to lunar eclipses on Earth, when from here we can see the Earth's shadow move across, and sometimes cover, the Moon.[3] If you are on the Moon during a lunar eclipse, you will see (through an appropriate filter) the Sun move behind the Earth. Just like in a beautiful sunset here on Earth, you will see the Earth's atmosphere take on a rusty hue.

Consider some of the other opportunities that you will have on the surface of the Moon:

Learn to Walk in Lower Gravity

Despite all your training on Earth, walking on the Moon will take some getting used to. You will weigh only about a sixth as much on the Moon as you do on Earth. Walking, running, jumping, reaching down, falling, and getting back up are actions our bodies are trained to deal with in normal Earth gravity. However, when we suddenly weigh much less, our muscles don't respond as they do normally. When orbiting the Earth and on your cruise to the Moon you will have experienced the extreme case of this muscle response issue in microgravity, especially with the challenges to your proprioception.

As noted in chapter 4, learning to walk on the Moon will be part of the training you receive before leaving Earth. If you trained underwater for it, you'll find that walking on the virtually airless Moon has a different "feel" than does walking or falling in water. By adding weight to you, the space suit you will be wearing will help you a little. The space suits worn by *Apollo* astronauts walking on the Moon weighed about 200 pounds (90 kg) on Earth. On the Moon, the suit itself weighed about 35 pounds (16 kg). So, if you weigh 175 pounds (80 kg) today, you will weigh about 30 pounds (14 kg) on the Moon without the suit, and 65 pounds (30 kg) with the suit. Even with all the training you will have had, your first Moon walks will take some adjustment.

Study the Law of Gravity

The Moon is a good place to learn some science. For example, most people believe that heavier objects fall faster than do lighter ones. Indeed, if you crumple a piece of paper and drop it and a pencil simultaneously from the same height here on Earth, as I just did, the pencil *will* strike the ground first. However, the reason that happens is not that gravity pulls the heavier pencil more strongly than the piece of paper, but because air friction slows the paper more than the pencil. Without air friction, the mass (see chapter 1) of an object, and hence its weight, has no effect on how fast it falls. To prove this, *Apollo 15* astronaut David Scott dropped a hammer and feather simultaneously from the same height on the airless Moon, and they reached the ground at the same time. You can see this on YouTube. You can also do it yourself on the Moon with *any* pair of objects available to you. Getting used to how gravity really works will enable you to take the next step, which is to try activities that you did on Earth, but that will be different on the Moon.

Play Sports

Many sports activities will have a very different "feel" to them on the Moon than they do on Earth. For example, a soccer ball on Earth weighs 1 pound (½ kg); hence it will weigh ⅙ pound (just under 3 ounces or 75 grams) on the Moon. Kicking a soccer ball there with the same force as you do on Earth would send it much higher and farther than it goes here. Likewise, golf on the Moon, first played by astronaut Alan Shepard in 1971, will give you a substantially better drive than on Earth. Golf courses set up there will have much longer fairways than courses on Earth. If you think finding your golf ball in the rough is tough on Earth, consider how quickly and completely your golf ball will become gray due to the static cling of regolith dust. Color won't help you find it on the Moon, so the manufacturers will likely insert tiny radio transponders in golf balls. You will be able to download an app on your tablet for tracking them.

In summary, every sport in which throwing, hitting, or kicking a ball is a factor, and in which the same ball is used there that is used here, will have to be relearned and played on much large grounds than the same sport on Earth. If balls and other sports equipment are made six times heavier than they are on Earth, you will be able to throw or hit them as you would today. They will still fall more slowly than they do here, so they will go farther.

Ride

As on Earth, you will travel by vehicle to different sites. It is possible that internal combustion engines (like car motors) could be designed and built for the Moon. Cars running on fuel will not use a petroleum product or natural gas, neither of which exists on the Moon. Rather, they will use liquid hydrogen, which will be "burned" with liquid oxygen. Both the hydrogen and the oxygen will come from the water found on the Moon, separated into its components, liquefied, stored separately, and pumped into vehicles at the equivalent of gas stations up there.

Electric vehicles for use on the Moon can be recharged from the solar electric grid that will be developed. That grid will have to pretty much encircle the Moon because during the 14¾ days of continuous darkness each month, supplies of electricity at any given site will likely be used up. Having solar cells distributed around the Moon and connected by wires, like power lines on Earth, will make electricity available everywhere it is needed at all times. Being able to ride many kilometers or miles will allow you to visit a variety of geological features on the Moon, maybe the biggest draw for going there.

Visit Geological Features

The International Astronomical Union, our umbrella international astronomical organization in charge of naming objects and features on objects in space, among other things, has identified eighteen categories of features on the Moon. The features originally put in one of these categories, fossa (long, narrow trenches), have been recategorized. Since there is overlap among a variety of the remaining features, I have combined them into the topics described below. *This list is intended to give you a sense of what geological features the Moon has and that you could visit, rather than a complete catalogue, which would take hundreds of pages.* By the time you go, specific trips to various of these places will be available, as well as the opportunity to explore previously unexplored regions.

Mare

The Moon has two fundamentally different types of surface regions, as shown in figure 9.4. About 16 percent of its surface is covered with relatively level, dark regolith called mare (plural: maria, Latin for seas, as they were once thought to be). The rest is covered with lighter gray highlands, which have mountains and many more impact craters than the maria. While astronomers and geologists are still debating the details of their formation, it appears that maria are composed of rock called basalts that seeped out as lava from deep inside the Moon well after that body first formed. This lava filled gigantic craters created earlier. As occurs on Earth, once the lava (called magma inside the Moon or Earth) flowed out, it began cooling and solidifying.

We learn about the history of the Moon by examining pieces of it brought back to Earth. The lunar mare basalts brought back by the lunar astronauts and the former Soviet Union's *Luna 16*, *Luna 20*, and *Luna 24* spacecraft are from 3.2 to 4.2 billion years old, whereas the rocks brought back from a more mountainous highland region adjacent to a mare are between 3.9 and 4.4 billion years old.

Because the maria are smoother than the other parts of the Moon, most landers have set down on them. This may not be necessary when you go. Since the highlands have a greater variety of geological features to visit, it is likely that some commercial landing sites will also be built there.

Rilles/Rima

The Moon appears to have dry riverbeds called rilles (German) or rima (Latin), but they were not created by water flow. As occurred on Earth, there were likely winding, underground rivers of molten, red-hot magma flowing through the young solidifying lava that was forming the maria. When these rivers cooled, they stopped flowing and solidified. The process of transforming from liquid to solid caused the magma to contract, since solid rock is denser and therefore takes up less space than the liquid rock from which it came. Due to this contraction, the underground rivers became tunnels. The surface basalt above these tunnels was solid rock, which could support the empty spaces created below it. However, over billions of years, particles from the Sun and from interstellar space pulverized this surface rock, creating the powdery regolith. As a result of the rock above the tunnels being turned to powder, the roofs weakened and eventually caved in, creating rilles, which look like dried riverbeds. It is likely that the collapses were not complete, meaning that there may still be caves in some of the rilles. There also may be caves where rilles have not yet formed above other solidified underground lava rivers.

Another plausible source of rilles is the flow of surface lava rivers that eventually solidified and, upon contracting, left these winding valleys. Astrogeologists are still debating whether both of these mechanisms caused rilles on the surface of the Moon. *Apollo 15* astronauts David Scott and James Irwin drove their Moon buggy to Hadley Rille (figure 9.5), named after the eighteenth-century British mathematician and inventor John Hadley (1682–1744). It is likely that you will be driven down into a rille to explore its properties, possibly including the natural tunnels that some of them are believed to still have.

FIGURE 9.5

Astronaut David Scott with the *Apollo 15* Lunar Rover at the edge of Hadley Rille in July 1971.

NASA

Scarps/Rupes

There are also cliffs on the Moon called scarps or rupes. As noted above, when a liquid solidifies (or a gas liquefies) it typically takes up less volume than it did before the transition. Therefore, as the molten interior of the Moon has been cooling and solidifying over billions of years, it has been shrinking. However, the surface became solid while the interior was still molten. As the core cools and solidifies, the interior rock becomes more compact, leaving a space above it. But the solid surface couldn't contract smoothly with it; some parts of the surface had to move in faster than others, leaving cliffs called scarps or rupes at the boundaries between regions that contracted faster or slower (figure 9.6). This is analogous to how fruits shrivel as they dry on the inside.

Over 80 scarps are known to exist. *Apollo 17* astronauts David Scott and James Irwin drove their moon buggy with great difficulty up over the intersection of the Lincoln and Lee scarps. The scarp walls you can visit are likely to have interesting geological features related to the early history of the Moon.

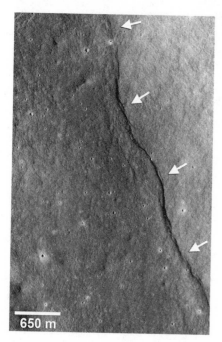

650 m

FIGURE 9.6

Scarp on the far side of the Moon.

NASA/Goddard/Arizona
State University/Smithsonian

Oceanus

The Mare Procellarum (Sea of Storms) is so large that it was given the name Oceanus Procellarum (Ocean of Storms; visible as the large, dark gray region to the left of center in figure 9.4). It is the only mare large enough to rate this title.

Paludes

Latin for swamps or marshes (singular: palus), three paludes, Palus Epidermiarum (Marsh of Epidemics), Palus Putredinis (Marsh of Decay), and Palus Somni (Marsh of Sleep) have been identified on the Moon. They are relatively flat regions that are smaller than maria (figure 9.7). The names

FIGURE 9.7

Image of the Apennine mountains, lower right, and the *Apollo 15* landing site (arrow), which was on the edge of the Palus Putredinis (Marsh of Decay).

NASA/University of Arizona

are *not* associated with known negative factors for these regions. Neither they nor any of the rest of the Moon's surface have any liquid water.

Planitia

Latin for plain, indicating a relatively small flat region. Planitia Descensus is the only officially designated planitia on the Moon (figure 9.8). In comparison, there are ten on Mars.

FIGURE 9.8

Planitia Descensus, at the top of this image, is a level, relatively crater-free region on the Moon.

NASA

Montes

The Moon has many mountains and mountain chains in its highlands. They are called montes (Latin, singular: mons). Unlike the mountains on Earth, which mostly formed as a result of tectonic plate motion over millions of years, mountains on the Moon formed within hours as a result of large pieces of space debris slamming into the surface. Among other things, those impacts pushed a lot of the outer layers around the impact sites aside, forcing them into surrounding material. Since all the material couldn't move sideways, some of it moved up, creating the mountains (figure 9.9).

You may be able to climb a mountain on the Moon, as many of them have relatively shallow slopes and, of course, you will weigh less than you do here, even with full space suit. It appears that the sides of most of the mountains are powdered regolith, like the surface of the maria. This may create problems on especially steep slopes, as the regolith would relatively easily slide downhill, creating an avalanche.

Craters

Craters will likely be a popular destination for travelers on the Moon. To the best of astronomers' knowledge, nearly all the craters on the Moon formed as a result of impacts by space debris, whereas most craters on Earth formed from volcanic activity. (Some remnants of small volcanoes have been reported on the Moon.)

There are several notable features of newly formed craters. The more massive an object, and/or the faster it is moving when it hits the Moon, the bigger the resulting crater. When the object strikes the Moon's surface, it splashes debris called ejecta out from the impact site. This is similar to what happens when you throw a rock into water here on Earth. As noted earlier, unless it falls straight down, water preferentially splashes in the direction the rock is moving. However, the speed of most incoming bodies on the Moon is so great that the resulting explosion sends ejecta equally in all directions, which is why impact craters there are circular.

The debris ejected on the Moon from the impact site is called its ejecta blanket. This ejecta is initially smoother and therefore brighter than the

FIGURE 9.9

Montes Apenninus mountain chain as taken for *Apollo 15*. The white dot is the *Apollo 15* landing site; Hadley Rille is visible below it. The mountains extend about 1.6 to 4.8 km (1 to 3 mi) above the landing site's basin.

NASA/JSC/Arizona State University

surrounding regolith, but it darkens with time as it is made rougher by impacts from the solar wind particles and cosmic rays. By observing how bright an ejecta blanket is, astronomers can estimate how long ago its crater formed. Eventually the ejecta blanket is as dark as the surrounding regolith and hence is no longer visible.

Another feature of each crater is the crater wall, made of debris pushed out by the explosive power of the impact. This is loosely compacted, powdery rock that can, and sometimes does, undergo landslides. Some crater walls are sufficiently weak that parts of them slide back down into the crater, such as when another piece of space debris lands nearby, causing a moonquake that shakes loose some of the regolith on the crater walls.

Some crater-forming impacts are so powerful that the central region of the crater is supercompressed and then springs back, creating a central peak or ring. These are called complex impact craters. Figure 9.10a summarizes these features. Figure 9.10b shows a wide distribution of craters on the Moon.

You may be able to choose which craters to visit. Whichever you choose, be careful not to slide down the crater wall, which could damage your space suit, or do something that creates a significant landslide of crater wall debris if someone is below you. However, you may be able to ride the equivalent of a snowboard down the side of especially steep craters. Whether this will be possible will depend in part on the stability of crater walls and on the viscosity of the regolith—how easy it is to slide over.

Promontorium

The boundaries between highlands and maria are often filled with interesting features. Indeed, a variety of mountainous regions extend out into the

FIGURE 9.10A

Crater Theophilus. This image was taken by a camera on *Apollo 16*, part of which is visible on the left. You can see the central peak, the crater wall (parts of which have slid into the crater) and the ejecta blanket, which is lighter colored than the surrounding surface.

NASA

FIGURE 9.10B

Crater field on the Moon imaged by the crew of *Apollo 11*.

NASA

maria. Similar features on Earth are called promontories, hence the name promintorium for such features on space bodies. Nine features have been designated as such on the Moon. Two of them, promontorium Heraclides and promontorium Laplace, are shown in figure 9.11.

Sinus

Known by the Latin term for a bent or curved surface, these features on the Moon often have the appearance of a bay, as seen on the coastlines

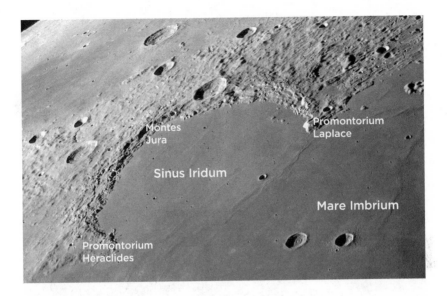

FIGURE 9.11

On the boundary of Mare Imbrium is the sinus Iridium, surrounded by montes Jura, which ends in the promontorium Heraclides and Laplace. Several notable craters are also visible.

NASA/GSFC/Arizona State University

of Earth. They are formed in impact craters. Lava from inside the Moon filled them up (see figure 9.11).

Landing Sites

For those interested in history, the existing landing sites of spacecraft from the twentieth and early twenty-first centuries will be a big draw. It is extremely likely that by the time you go to the Moon, the *Apollo* landing sites (figure 9.12) and the former Soviet Union's Lunokhod (Russian for Moon walker) rover sites will be declared international landmarks, from which it will be strictly forbidden to remove memorabilia or to tread on early astronaut boot prints. Another possible option for this historic material on the Moon is to bring it all back, put it in museums, and replace it with replicas, which will still be off limits to visitors. In any event, guided tours of these locations will likely be available.

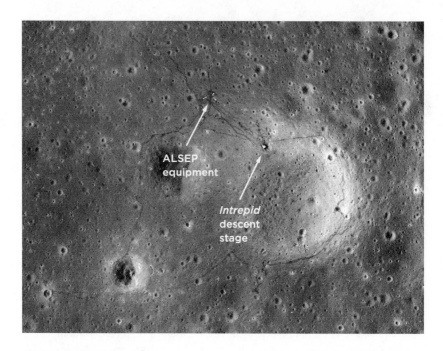

FIGURE 9.12A

Apollo 12 landing site, with rover tracks and footprints. ALSEP is short for Apollo Lunar Surface Experiment Package.

NASA/Goddard Space Flight Center/ASU

Lacus

Latin for lake. These are depressions in the surface of the Moon that, if there was liquid water in them, we would call lakes (figure 9.13). There are presently twenty such regions identified on the Moon, and others on other worlds.

Dorsa

The maria we saw above have a variety of structures. Among these are elongated mounds called dorsa (singular: dorsum; Latin for the back of a body). Poetically called wrinkle ridges, they often occur in parallel groups that can run for hundreds of kilometers or miles (figure 9.14). They formed

AS12-47-6921

FIGURE 9.12B

Astronaut Charles Conrad Jr. setting up parts of ALSEP on *Apollo 12*.

inside all the maria as this lava cooled and contracted, leaving some debris higher than its surrounding surface.

Albedo Feature

Some places on the Moon are unusually bright. They are said to have high albedos. Albedo is the percent or fraction of incident light that is scattered directly back into space from the surface or upper cloud layer of any astronomical object. Albedo features are regions on the surface of the Moon in which the amount of light scattered is much different from the surrounding area. There is one on the near side, Reiner Gamma (figure 9.15), and several on the far side. Reiner Gamma is particularly interesting because the brighter areas have a swirling shape. The cause of albedo features is unknown, but they may be due to local magnetic fields or energy focused

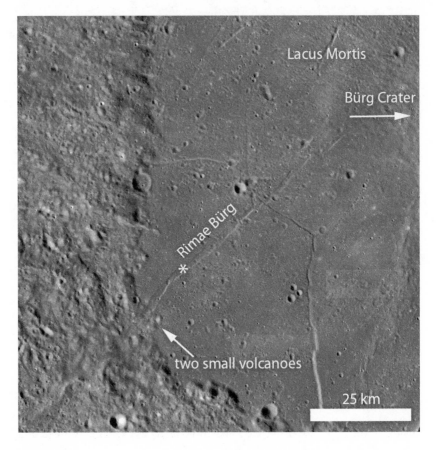

FIGURE 9.13

Part of Lacus Mortis. Note the rilles running through this image.

NASA

there from an impact on the other side of the Moon. The results of such focused energy from impacts are observed on the planet Mercury.

Catena

While most craters on the Moon are distributed randomly over its surface, there are some exceptions. Catena describes a chain of craters in a nearly straight line that were formed when a set of debris, such as pieces of a comet that had come apart and were traveling in the same orbit, struck

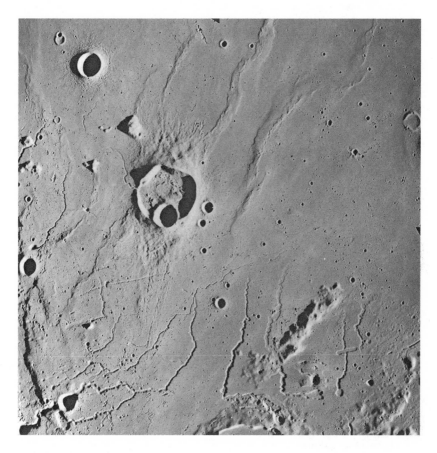

FIGURE 9.14

Apollo 15 image of Kreiger crater (slightly above and to the left of center) with rilles below it and wrinkle ridges above it.

NASA/JSC/Arizona State University

the Moon (figure 9.16). Other explanations are also under investigation. To date, twenty catena have been found on the Moon, some on the near side, others on the far side.

Vallis

Named with the Latin term for valley, these are valleys that apparently formed in a variety of ways. Some came from the nonuniform descent

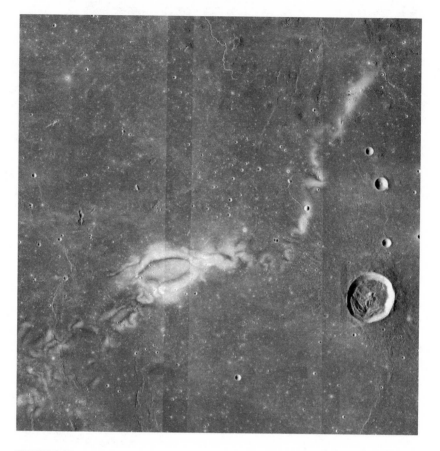

FIGURE 9.15

Albedo feature Reiner Gamma, an especially bright region on the near side of the Moon. The cause of this formation and properties of the material there are still under investigation.

NASA LRO WAC science team

of the surface (figure 9.17). Some were formed from a string of adjacent craters. Some radiate from large craters and probably formed from the impacts that created them.

Visit Natural Openings and Tunnels in the Moon's Surface

There may be a variety of tunnels near the surface of the Moon. As noted above, astrogeologists believe that at least some of the Moon's rilles were

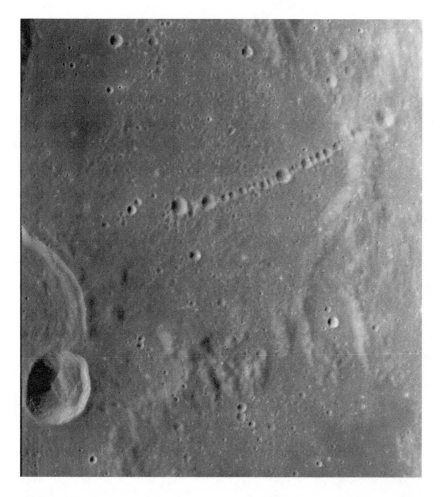

FIGURE 9.16

Catena Davy, a string of impact craters in the Mare Nubium presumably created at the same time, possibly from the debris of a comet that had come apart.

NASA/GSFC/Arizona State University

formed by collapsing underground lava rivers, some of which may contain caves. If any of these tunnels, called lava tubes, exist, you will be able to explore them just as you can explore lava tubes on Earth, notably in Hawaii, Iceland, Spain, Portugal, and various places in North America.

Astronomers are also discovering openings in the surface of the Moon (figure 9.18). They may be the roofs of lava tubes that have collapsed, as

FIGURE 9.17

Vallis Alpes, valley on the Moon. Notice also maria, mountains, craters, and rilles.

NASA/USGS/LPI

also occurs on Earth, leaving similar openings, or may have different origins. Where they lead is unknown, but if your hotel is close enough, you will be able to visit one of these openings and possibly explore any tunnel there. Indeed, your hotel could be in one of these caves.

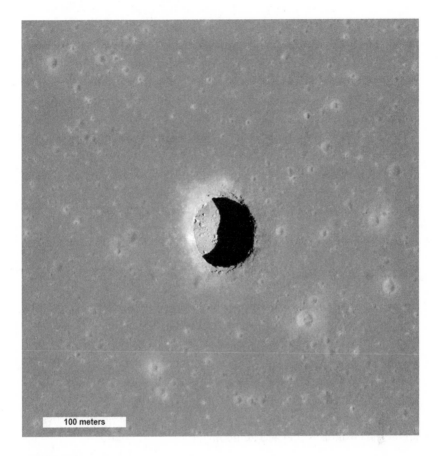

100 meters

FIGURE 9.18

Opening on Mare Tranquilitatis. Whether this goes into a lava tube or another cave structure on the Moon is presently unknown.

NASA/GSFC/Arizona State University

Use Telescopes

Besides visiting the physical and historical sites on the Moon, you can study astronomical objects (including the Earth) from there through a powerful telescope. There are already plans to put research-grade telescopes on the far side of the Moon. The total lack of lunar atmosphere will allow these telescopes to be used 24/7. Without atmosphere to scatter its light, sunlight can't get into the telescope unless the instrument is aimed directly at it. The far side of the Moon is also ideal for radio telescopes, which normally have to deal with interference from other radio sources on Earth, such as cell phones and radio stations. The Moon absorbs that radiation from Earth, preventing it from getting to the far side.

Putting observatories on the near side of the Moon would not be as efficient as putting them on the far side because the Earth would be a large bright spot, as seen from the Moon. Furthermore, the Earth's atmosphere gives off light that would enlarge the area toward which the telescope could not aim. Nevertheless, it would be a draw for hotels and habitats on the near side of the Moon to have intermediate-sized telescopes for their guests and inhabitants to use. In the trade we call these 1-meter class telescopes, meaning that the large, light-gathering mirror on them is about 1 meter or 1 yard across. The magnification from such telescopes gives absolutely breathtaking images of millions of interesting astronomical objects. Furthermore, the lack of atmosphere on the Moon will prevent light entering the telescope from twinkling, as discussed earlier, making images much sharper than most seen from Earth.

A visit to a professional observatory on the Moon, like visits to such observatories on Earth, will give you an unparalleled perspective on research astronomy. Since the telescopes at these facilities will be operating continually and the images they collect can be shown on monitors practically in real time, you may have the opportunity to see new discoveries about the cosmos as they are being made.

Visit Water Mining Operations

As the Moon is developed, mining and manufacturing facilities are likely to be built there. Perhaps one of the first will be for the recovery of water. As barren as the Moon appears, astronomers have discovered that there are water ice and water-related compounds under its surface in many places. Ice has been discovered in many craters near the Moon's poles that are never exposed to sunlight. Estimates are that there are cubic miles of water ice there, and perhaps equivalent amounts of molecules that can be combined to make water located all over the Moon's surface, probably delivered by ice-rich comet and asteroid debris. There appears to be enough to make recovering it economically viable.

You may be surprised that water in space has more uses than it does on Earth! All of the nontraditional uses begin with splitting the water into its constituent hydrogen, H_2, and oxygen, O_2, molecules. This is done by a process called electrolysis, using electrical energy from solar panels. The resulting hydrogen and oxygen molecules are stored in separate tanks, as noted earlier.

Some of the oxygen can be used for breathing.

Some of the hydrogen and oxygen can be recombined and used for heating and making electricity. When these gases are combined and ignited, they create heat and hot water vapor (i.e., steam). This can be done under very controlled conditions on the Moon in the equivalent of a furnace on Earth. The heat so released can be used to heat habitats during lunar nights or to generate electricity or move vehicles. The water, when cooled, can be used for drinking and other "normal" purposes.

Some of the hydrogen and oxygen can also be used to make rocket fuel. Combining and igniting the hydrogen and oxygen gases in large quantities and letting the resulting very hot gases come out in controlled streams would create a rocket. Therefore, the hydrogen and oxygen made from water there can be used as the rocket fuel for spacecraft landing on and leaving the Moon, or to fuel other spacecraft waiting in orbit around it.

Collect Samples of Moon Rock to Bring Home

One of the major concerns in going into space *from* Earth is the weight (or mass) of each rocket and its cargo. The higher that total weight, the more rocket fuel needed to put it into orbit. So critical is this issue that everything going up from Earth is measured to the last ounce or gram. Because of the lower weight of objects on the Moon, it will be economically feasible to make rockets that land on and take off from the Moon capable of carrying considerable mass up into orbit around it, and also to make Earth-Moon shuttles that can take the Moon rocks you collect back to Earth via the Earth-orbiting space station. Therefore, it is likely that each lunar traveler will be able to bring back at least several pounds or kilograms of Moon rocks, which is why you will be given the opportunity to visit rock-rich areas on the Moon to collect them.

While the geology of the Moon rocks brought back to Earth by the *Apollo* astronauts is well understood, these rocks represent a very small sample of what lunar geologists expect to eventually discover. By the time you go to the Moon, the options will be better understood, and it would be worth your time and effort to learn what are the most attractive and valuable rocks to collect. Keep in mind too that discoveries of valuable minerals on Earth, such as diamonds in South Africa in 1867, are often accidental. You may stumble across a vein of a unique mineral during your visit to the Moon.

Observe Lunar Transient Phenomena

For hundreds of years, people have reported seeing short-lived changes in some features on the Moon. Most of these events are reported as bright flashes or bursts of color. The impacts of space debris, discussed in chapter 7, are likely to have caused some of these. The causes of the rest of them, if they really occur, are presently unknown. The problem in determining the validity of most of these reports is that they are so short-lived that usually only one person sees each event. Traditionally called lunar transient phenomena, they are now also called transient lunar phenomena. One particularly intriguing report came when *Apollo 11* astronauts were on the Moon. Here is part of the transcript from NASA:

> **BRUCE MCCANDLESS (ON EARTH):** Roger. And we've got an observation you can make if you have some time up there. There's been some lunar transient events reported in the vicinity of Aristarchus. Over.
>
> **NEIL ARMSTRONG:** We're coming up on Aristarchus right now.
>
> **MICHAEL COLLINS:** Hey, Houston. I'm looking north up toward Aristarchus now, and I can't really tell at that distance whether I am really looking at Aristarchus, but there's an area that is considerably more illuminated than the surrounding area. It just has—seems to have a slight amount of fluorescence to it. A crater can be seen, and the area around the crater is quite bright.
>
> **BUZZ ALDRIN:** Houston, *Apollo 11*. Looking up at the same area now and it does seem to be reflecting some of the earthshine. I'm not sure whether it was worked out to be about zero phase to—Well, at least there is

one wall of the crater that seems to be more illuminated than the others, and that one—if we are lining up with the Earth correctly—does seem to put it about at zero phase. That area is definitely lighter than anything else that I could see out this window. I am not sure that I am really identifying any phosphorescence, but that definitely is lighter than anything else in the neighborhood.

The discussion continued for about half an hour.

If lunar transient phenomena are really occurring, and my bet is that they are, a variety of events could be causing them. Besides debris kicked up by small impacts on the Moon, these include gases escaping from inside the Moon and electrostatic charges (static electricity) setting some gases in motion when the Moon is struck by high-energy particles from the Sun. Knowing that they may occur, you are in a better position to look for these either directly as they occur or in live feeds moments after they occur. The more lunar transient phenomena that are observed by several people or more, the better scientists will be able to get a handle on what they are and what causes them.

Sex

Sex on the Moon will be a different experience than sex in microgravity, primarily because you can use basically the same "mechanics" as you do on Earth, whereas everything is different in weightlessness. Remember that on the Moon both you and your partner will each weigh a sixth as much as you do on Earth. Therefore, you will have the opportunity to engage in sex more comfortably and in many more positions than you can on Earth.

JOURNEYS TO ASTEROIDS/COMETS/MARS MOONS

When you leave either Earth orbit or lunar orbit for more distant astronomical destinations, you will begin a trip of epic proportions. However, the acceleration you will undergo during departure from our neighborhood is much less than you experienced leaving Earth. As discussed in chapter 2, your path outward is likely to be a modified Hohmann orbit, which is energy efficient and relatively fast. In any event, however, we are talking about months in transit, so a discussion of making the most of this travel time is important.

Long-term (long-distance) space voyages will be filled with a variety of everyday activities, many of which were introduced earlier in the book and are listed here:

- eating
- exercising
- applying liquid makeup, doing hair, shaving
- going to the bathroom
- reviewing in-flight procedures and rehearsing for normal and emergency situations
- communicating with friends and family back home
- blogging and tweeting
- journaling
- taking classes, either from fellow travelers or from Earth
- doing research in your professional field or areas of interest
- keeping up with news from Earth
- participating in business back home, if you are so inclined
- sex
- working
- socializing

Some of these activities deserve further comment. Communicating with friends and family will be interesting. It is wonderful to get good news and to follow family activities and family adventures. Today, you can Skype or otherwise communicate face to face. However, since all electromagnetic radiation travels at the same finite speed, the farther you get from Earth, the longer it takes for messages from Earth to get to you or from you back to Earth. For example, while it only takes about one second for signals to travel to the Moon from Earth or back, when you are at a Trojan asteroid at Earth's Lagrange point, it takes 4½ minutes each way. After each comment you make into the monitor, it will take at least 9 minutes to get a response. You'll have time to get a cup of coffee between, "Hi! How are you?" and "I'm fine, but the cat is sick."

SCIENCE AND SCIENCE FICTION

In 1905 Albert Einstein published his "special theory of relativity," which explained mathematically why nothing can travel through space faster than the speed of light. "Light," as used here, means any kind of electromagnetic radiation. Indeed, all types of electromagnetic radiation (radio waves, microwaves, infrared radiation, visible light, ultraviolet radiation, X-rays, and gamma rays) travel at exactly the same speed. The reason matter can't travel faster than light is that as it speeds up, all matter actually becomes more massive. For example, if you sent a 1 pound (½ kg) object into space at 87 percent the speed of light, its mass would increase to 2 pounds (1 kg) as measured from your starting point. You couldn't send it out at the speed of light, because at that speed it would have more mass than everything in the entire universe!

To date, every prediction of special relativity has been verified by experiment. Furthermore, there is no experimental evidence for another dimension that would allow for "hyperdrive" speeds faster than light. The speed of light is the limit—no warp factor for you.

The radio signal time from Earth to Mars (or its moons) and back varies with the relative distance between Earth and Mars in their orbits around the Sun. The shortest one-way signal time is about 3 minutes, when the planets are nearest each other on the same side of the Sun, while the longest one-way signal time is 22¼ minutes, when they are on directly opposite sides of the Sun. This time delay will likely lead to a new style of communication in which each party simultaneously starts a separate story line, cutting the "dead air" time in half.

Millions of people back home will be interested in communicating with you during your voyage. You may be able to parlay that into a learning experience for them, for which you are paid. Conversely, if there are topics you have always wanted to learn about but haven't had the time to study, your trip might be a great opportunity. Teaching or taking courses are just two examples of an essential theme: enjoying your transit times in space. As mentioned earlier, experience with sailors in submarines, people living in isolated environments, and astronauts on long-term orbital missions reveals that everyone feels much better about those experiences if they truly believe they are doing something useful. Therefore, your travel company will work with you prior to departure to put together a very constructive plan for the use of your time.

This brings up the issue of cycles in your life on a spaceship. We all have cycles of activity in our lives. As discussed in chapter 7 and above under the topic of space station visits, we humans evolved to live in a twenty-four-hour sleep/wake cycle. As on orbiting space stations, the lighting, temperature, and sound in your spaceship will have twenty-four-hour cycles so that your body's circadian rhythm will not be disrupted.

There have been experiments to see what the best length of the week is for our comfort. For example, NASA tried having astronauts function on a ten-day week, working for eight days, with two days off, but the astronauts didn't like it. As on Earth, your spaceship will function on a seven-day week, with five to six days focused on constructive activities and one or two days primarily for recreation.

The issue of recreation is the tip of a very important iceberg to be considered for enjoying your time on a long-haul voyage in space. As introduced in chapter 8, groups in isolated environments very often form cliques in which other members of the ensemble are not welcome. The whole issue of group dynamics in space travel is extremely important, not just for recreation but for all aspects of the journey. In fact, other than doing productive work and feeling good about it, maintaining harmony may become the paramount issue for long space trips.

Recreational activities with your fellow travelers will be available and encouraged on long trips, and other long-term activities, such as online gaming of all sorts, will help you while away the recreational hours.

■ ■ ■

Visiting different types of worlds will require careful attention to their chemical composition. Recall from chapter 1 that asteroids are composed primarily of rock and metal, with varying amounts of ices. Comets are composed primarily of rock and ice. The moons of Mars, Phobos and

Deimos, appear to be captured asteroids; hence they are likely to be mostly rock and metal. Recall also that comets emit considerable amounts of ice and rocky debris when they pass close to the Sun. Much of this material is being pushed away from the Sun, thereby creating comet tails. It would likely be quite dangerous to fly into the dust tail of a comet departing from the Sun, as some of that material could impact the spacecraft and thereby damage it. Comets approaching the Sun from the outer solar system, however, have not been heated as much as the departing comets and are much less active. Therefore, they are the ones that will be destinations for space travelers in the near future.

Because of their complex orbits, you may be the first visitors to such destination worlds. Being among the first visitors to any world is likely to be among the most satisfying experiences imaginable, like discovering a long-lost tomb or being elected president. Watching your destination grow larger on the monitor, just as we saw the dwarf planet Pluto in detail for the first time as I wrote this book, will be unforgettable. Although you will send images back to Earth, *you are really there!*

You will be able to look at all parts of your destination right out the window as you orbit around it. Indeed, the *Rosetta* spacecraft was put into orbit around comet 67P/Churyumov-Gerasimenko on September 10, 2014, on the comet's way toward the Sun, enabling us to see all parts of it. In addition, when you are safely in orbit around your destination world, you will be able to exit your spacecraft and land on it. This can be somewhat challenging if the object is spinning rapidly, but landing on such bodies is well within our technological capability. For all asteroids, comets, and the moons of Mars, their masses are so tiny that even when you land on them, you will be nearly weightless. For example, standing on the moon Phobos, you would weigh only slightly more than 1/2000 as much as you weigh now. I would weigh about .1 pound (0.05 kg) there.

One fun thing that you will be able to do from virtually any of these worlds is to jump off and float away. Although their gravitational attraction would continually pull back on you, if you launched yourself fast enough (faster than what is called the escape velocity of that world), you would actually keep on going, never to return. However, have no fear! You will be securely attached to your destination world by a tether, which will be sufficiently long so that you can enjoy jumping off it.

Walking on these low-gravity worlds will be a challenge. A network of horizontal cables will be put in place over the world you visit so that you can walk around and explore virtually every nook and cranny on it. Who knows what you will find? As on the Moon, anticipate becoming covered with dirt, held to your space suit by static electricity. You will be able to bring souvenir rocks and dust back home with you.

If your destination world is a comet, you may encounter jets of gas emerging from inside it. These will be dangerous because they likely carry outward small pieces of debris, as noted earlier. Nevertheless, they will be interesting to see. If your travel company brings along sheets of plastic to place over a jet, you can use them to collect specimens of the gases and larger debris being ejected into space. Note that you won't hear sounds coming from these jets, since all of these worlds—comets, asteroids, and the moons of Mars—have virtually no atmosphere to carry sound. Even the comas of comets are exceptionally thin atmospheres.

Moons of Mars

Visiting the moons of Mars will provide interesting perspectives about the Red Planet. Like our Moon around Earth, Phobos and Deimos are in synchronous rotation around Mars. As discussed earlier in this chapter, this means they each rotate at the same rate that they orbit, so that the same side of each moon is fixed facing Mars at all times. To make the most of your time on the moons, you will land on the side of Phobos or Deimos (you will visit both worlds) facing Mars. This will allow you to spend a considerable amount of time viewing the planet, as well as exploring the moon. These moons orbit Mars so fast that you will be able to view virtually the entire planet during your stay on its moons.

The orbits of Phobos and Deimos are very different from the orbit of our Moon. Both moons are many times closer to Mars than our Moon is to the Earth; Phobos is about 6,000 km (3,700 mi) from the surface of Mars and Deimos is about 20,000 km (12,400 mi) from the surface, while our Moon is typically about 384,000 km (238,000 mi) from Earth's surface. Therefore, Mars will appear to fill much more of the sky from either of its moons than Earth does, as seen from our Moon. Specifically, from our Moon, the disk of Earth spans an angle of about 2° in the sky, while from Phobos, Mars spans an angle of about 43° and from Deimos about 16½°. For comparison, our Moon spreads only ½° across the sky as seen from Earth.

Phobos is the larger and more heavily cratered of the two moons. The diameter of the largest crater, Stickney, is half the diameter of that entire moon. It is on the Mars-facing side of Phobos and is a must-visit feature of that moon (see figure 2.4). Named after Angeline Stickney-Hall (1830–1892), the wife of Asaph Hall (1829–1907), the astronomer who discovered the two moons in 1877, this crater was created by an impact that nearly destroyed that moon. Had that happened, some of the original moon would have been blown onto Mars, some of it would have been sent out of orbit forever, and the rest would have formed a ring around Mars. Bits and pieces of that ring, pulled by their mutual gravitational attraction, would have drifted together until eventually a new, smaller moon would have formed.

You will be able to visit regions of Phobos that are apparently beginning to come apart. Besides its craters, there are parallel valleys or grooves over much of this moon. They were originally thought to have been formed by the impact that created crater Stickney, but the latest proposal is that they were (and are) being created by the strains in the moon as it spirals in toward Mars. Just like our Moon's effects on Earth, Phobos is creating land tides on Mars. That moon is orbiting so rapidly that the tides pull back on it, causing it to lose energy and thereby spiral down toward Mars.

At its present distance, the gravitational force from Mars on the side of Phobos closest to it is measurably greater than is the gravitational force from Mars on the side farthest away. This difference in gravitational pull, called a tidal effect, is pulling Phobos apart, which is believed to be shown by the parallel valleys.

The moon Deimos (see figure 2.4) is more gently cratered, with no feature similar to Stickney on Phobos. Both moons have substantial regoliths, although we do not know how deep they are. With practice (or a specially designed spring-loaded launcher) and with a sufficiently long tether, you will be able to jump up hundreds or even thousands of meters (yards), stop due to the moon's gravitational field (as we are stopped by Earth's gravitational field when we jump up just a few centimeters [inches] on Earth) and fall back down. Even falling distances of hundreds or thousands of meters toward either moon, you will land extremely slowly, at the same speed you left it going up. As on any other distant worlds, it will be well worth your time and effort to collect rocks and regolith to bring home.

We have the technology today to visit any of the worlds explored in this section. In regard to the moons, the question naturally arises: "If I'm going to the moons of Mars, why not visit the planet as well?" The short answer is that you *could* go if you paid for it, but the additional cost would be significant. The reason, explained in chapter 2, is twofold. First, it is extremely difficult to land on Mars. The planet's relatively strong gravitational attraction, compared to that of our Moon or any other object you can visit in space, combined with its very thin atmosphere, requires expensive technology for landing humans there safely. Second, developing spacecraft technology that can withstand the landing in perfect working condition so that at least part of the lander can blast back off into orbit is *very* challenging. Conversely, however, if you pay for a round-trip to Mars, then you can expect to visit its moons for a nominal extra fee.

VISITING MARS

Mars is the ultimate space destination in the near future. The journey for a visit to the planet's surface will follow the same route from Earth orbit as for a visit to its moons. Indeed, some passengers on a flight out might be going to the moons while the rest are going to the planet's surface. In any event, the rehearsals that each group would undergo during the last month of the trip would be very different. Moon walkers would go through tethering and low-gravity training, which visitors to the planet's surface would not have to do since, as we saw in chapter 2, the surface gravity of Mars is about .4 times the Earth's surface gravity. That means you would weigh .4 times as much there as on Earth. Instead, training for visitors to Mars will deal primarily with the landing and moving into the long-term habitats, which, like those on the Moon, will probably be underground to protect them and their occupants from exposure to the high radiation levels on the planet's surface. As on the Moon, your space suit and everything else exposed to the regolith will become dirty from static cling.

As noted earlier, the reason that countries and companies involved in space exploration are considering developing permanent habitats on Mars is that the expense of building and positioning rockets on Mars to take people off the planet is, pardon the expression, astronomical. Allowing that some people will be able to afford the extra tens of millions of U.S. dollars for a flight off Mars, what follows will explore opportunities for both visitors and settlers.

Perhaps the most important issue concerning human visits to Mars is whether life exists there. While evidence from surface features on Mars shows that liquid underground water likely exists, except for one possible fossil found in a meteorite ejected from Mars, we have not yet found any evidence that life exists there now or ever did. This is not an assertion that Mars hasn't ever supported life. It is just a statement based on the current scientific evidence.

And yet, there is virtually incontrovertible evidence that large quantities of surface liquid water once existed on Mars. On Earth, surface (and even underground) liquid water harbors myriad life forms. Since that is so here, we might reasonably expect that the same applied to Mars back in its early days. There are arguments pro and con, given its different mass than Earth, different distance from the Sun, different surface chemical composition, different atmospheric composition, lack of a substantial moon like ours, and different history of tectonic plate motion (none now) compared to Earth, among other things. Discovery of life on Mars, even microbial life, would have a profound impact on our perspective on the likelihood of life elsewhere in the universe, as well as how life forms in the first place. While underground life on Mars is an entirely open question, the lack of liquid surface water there, the low atmospheric density and lack of oxygen, and the extremely high radiation levels from the Sun combine to make it exceedingly unlikely that life exists on the surface today.

If there is life in the liquid underground water, the bacterial component of it will be a major concern for the safety of humans there. If humans get any of those bacteria in their blood, we may not have the built-in biological protection necessary to kill them before they harm us. Bringing back potential contagions from Mars could have devastating effects for life on Earth. Conversely, while pharmaceutical laboratories might be able to develop vaccines to protect humans on Mars, bacterial life on Mars may not be able to survive in an environment contaminated with human bacteria and viruses. If this is true, our presence there could wipe out native life on Mars. Simple as that life may be, this does raise ethical questions about our being there. Such issues are under the purview of NASA's Office of Planetary Protection.

Air

There is virtually no breathable oxygen in Mars's thin atmosphere, which overall is only about 0.6 percent as dense as the air we breathe. The Martian atmosphere is composed almost entirely of carbon dioxide, argon, and nitrogen. While oxygen comprises 21 percent of the molecules in Earth's atmosphere, it is only 0.14 percent of the molecules in Mars's atmosphere. Therefore, you will have to carry bottled oxygen whenever you are out of a habitat. Likewise, the low air pressure will require that all habitats and vehicles designed to have breathable atmospheres be sealed so that the air can't leak out.

The thinness of the atmosphere also limits the amount of ozone (three oxygen atoms bounded together, O_3) that it contains, compared to the Earth. Our ozone layer blocks some of the more deadly ultraviolet radiation coming from space, so with its lower concentration of ozone, the surface of Mars experiences a far greater bath of potentially lethal solar radiation than we have here. Consequently, when you are outside on Mars, you will have to wear a space suit that shields you from this radiation.

Despite its thinness, the atmosphere of Mars supports water ice clouds, primarily in the low-lying equatorial regions and near the poles. These clouds are usually short-lived, often appearing and disappearing within a day or so. Snow has been reported falling from some of them.

Surface Features to Visit

Let us proceed by assuming that if life already exists on Mars, we can coexist with it, with neither species harming the other. Visitors are likely to stay for weeks or months, so let's assume that your travel company has arranged for you to visit all the major features and sites on Mars. Therefore, we begin with an overview of the planet's surface (figure 9.19).

The northern and southern hemispheres of Mars are very different environments. Much of the northern region is relatively smooth and crater-free, with a polar ice cap. Water ice is there as permafrost, while carbon dioxide ice (dry ice) varies seasonally. This smooth region of Mars extends in some places to the equator, while in other places it goes from the pole down about halfway to the equator. Astronomers believe that when Mars was young, before its surface water all froze or evaporated into space, this may have been a huge liquid water ocean.

In contrast, the southern hemisphere, with regions extending into the north, is much more heavily cratered, with several extinct volcanoes, a valley system that more than rivals the Grand Canyon in Arizona, dried

FIGURE 9.19

Unfolded surface map of Mars. North is at the top. The three mountains in a row on the left are the Tharsis Montes. From bottom to top, they are Ascraes Mons, Pavonas Mons, and Arsia Mons. The largest mountain on Mars, to the left of the Tharris Montes, is Olympus Mons. To the right of them is Valles Marineris.

National Geographic Society, MOLA Science Team, MSS, JPL, NASA riverbeds, vast plains, and a polar ice cap similar to that in the north. Overall, the southern hemisphere averages between 1.6 and 5 km (1 and 3 mi) higher than the northern hemisphere.

Three notable features appear along the boundary region between the two hemispheres. First, a variety of places have what is called fretted terrain (figure 9.20). This is very complex land with valleys, buttes, and mesas (flat-topped hills with steep sides), normal hills, and flows of debris. Astrogeologists haven't agreed on all the mechanisms involved, but some of it is apparently caused by processes including water flow, underground ice evaporation, winds, and glacier flows.

The second feature between the lowlands and the highlands is a complex multifaceted region called medusae fossae (figure 9.21). It runs along the Martian equator for about 965 km (600 mi) on the boundary between lowlands to the north and higher terrain to the south. It is composed of ash that is easily blown and hence resculpted by the wind, as well as ice deposits. Its origin is not yet known, but it shows a variety of features that

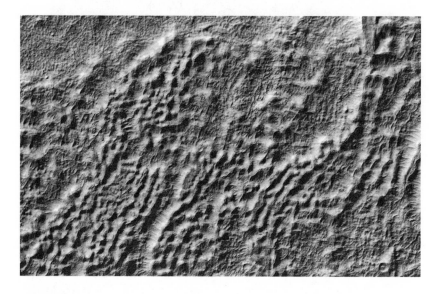

FIGURE 9.20

Fretted terrain on Mars.

NASA/JPL-Caltech/University of Arizona

suggest that impacts and water flow played roles in the formation process. It presents beautiful and exotic views, two of which are shown in figure 9.21a and b.

The third common feature in the mid-latitudes of Mars is called scalloped topography (figure 9.22). As the name suggests, the boundaries of these regions look like the edge of a scallop shell. The rounded edges cut into higher territory, as you can see in the figure. It is believed that these features are caused by the sublimation (evaporation directly from solid to gas) of water below them, not far from the planet's surface. The details of this process are still being worked out.

Besides potentially providing excellent places to visit, these features all strongly suggest that there is frozen water in very large quantities in equatorial regions of Mars. Indeed, the more astronomers study the entire

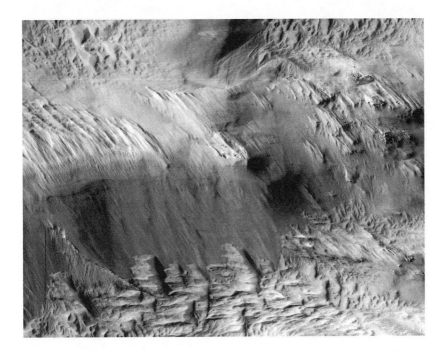

FIGURE 9.21A

Outcropping in the Medusae Fossae. The long hills are called yardangs.

NASA/JPL/University of Arizona

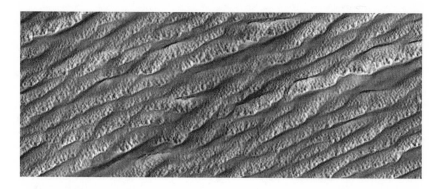

FIGURE 9.21B

Wind erosion of the Medusae Fossae.

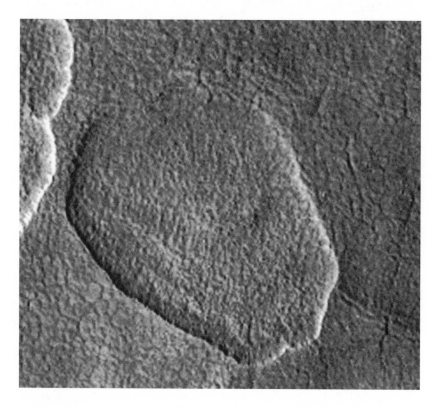

FIGURE 9.22

Scalloped topography on Utopia Planitia on Mars. Parts of the edges of the craters are scalloped.

surface of Mars, the more it appears that water has had an overwhelming impact. Its presence near the equator will make it that much easier to establish visiting sites and permanent living colonies in these warmest regions on Mars, compared to what would have to be done if the ices were only in the polar regions. This would require an extensive (read expensive) transportation network from the poles to the equator.

Another, more compact area on the boundary between the lowlands and the highlands is a region called Cydonia, filled with a variety of features that resemble pyramids, faces, skulls, and other objects made of rock, sand, and other regolith materials that at first glance suggested that advanced life existed on Mars at one time. Indeed, the more our rovers and orbiters photograph the surface of Mars, the more features they find that if seen on Earth, we would associate with humans or other life forms.

Before examining Cydonia in detail, it is worth commenting about this issue of advanced life. William of Occam (c. 1280–c. 1349) was an English philosopher who, among other things, is credited with proposing a crucial logical rule for selecting the most likely explanation of an object or event when there is more than one plausible explanation. Known as Occam's razor, the principle can be stated: *When several competing theories describe a concept with the same accuracy, choose as correct the one with the fewest unproven assumptions.* It is accepted as a guiding light by all scientists.[4]

In the case of the features on Mars mentioned above, there are two plausible explanations. First, perhaps they were created by some advanced life form, either for their own use or as messages to advanced life on Earth and elsewhere. Second, they may be the results of natural geological activity. According to Occam's razor, until proven otherwise, the geological explanation of all these features is to be held as the correct one. Briefly, different types of rock came to exist near the surface of Mars, either from inside or from impacts. The different types of rock then evolved (weathered) at different rates because they have different physical and chemical properties, allowing the recognizable features to form over billions of years. This theory does not require the unproven existence of advanced life that we have never seen, either evolving on Mars (incredibly unlikely, given the short amount of time that Mars had liquid surface water and a sustainable atmosphere) or stopping off their on a visit from another star system (i.e., interstellar aliens).

One might reasonably argue that if weathering of different surfaces creates such unusual features on Mars, then they should also occur on Earth. In fact, they do. These include columns of sedimentary rock called hoodoos; arches of rocks; salt columns in Mono Lake, California; mushroom-shaped chalk rocks in Egypt; and a column called Devil's Tower towering over the surrounding land in Wyoming, among many others.

An important aspect of how our brains work comes into play here. We are hardwired to see patterns and symmetries. You are much more likely to take note of parallel lines or a circle than a random squiggle on a page. Likewise, when you see objects that are similar in appearance to others, you are likely to connect them. For example, if you were driving through a desert in Nevada and saw a rock with this shape protruding a meter (a few feet) above the surrounding sand, it is likely that the first thought you'd have would be, "Oh my goodness, I've discovered the top of a pyramid." While that is possible, it is much more likely that you are seeing the top of a piece of metamorphic rock that has not weathered as much as the surrounding land, which is becoming sand and wearing away around it.

Objects ranging from a few centimeters (a few inches) to a few kilometers (a few miles) across that look like things we know on Earth have been observed on Mars. They include pyramids, skeletons, human faces, heart shapes, animal shapes, and a "happy face." Using Ockham's razor, all scientists believe that these are all rock formations that weathered naturally. These features represent a very tiny percentage of the objects visible on Mars, and given the extraordinary number of random shapes seen there, it is not at all surprising that some look recognizable.

Let's now consider Cydonia and some of the other parts of Mars you can explore.

Cydonia

Cydonia (figure 9.23) is particularly rich in features created by weathering and possibly by water erosion of different types of rock, combined with impact craters and debris sprayed out by these events. This region will be among the most attractive to visitors on the planet.

Valles Marineris

The canyon system of Mars, called Valles Marineris, in honor of the *Mariner 9* spacecraft that discovered it in 1972, is the most awe-inspiring feature on the Red Planet (figure 9.24). It is over 4,000 km (2,500 mi) long (the

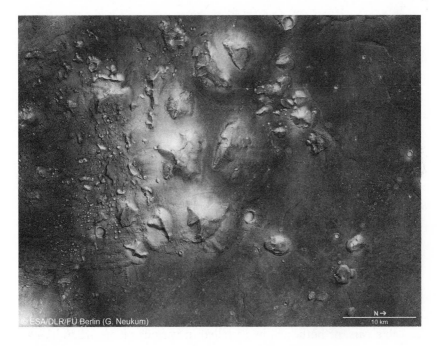

FIGURE 9.23

Cydonia, on Mars. How many features can you recognize? Hint: you can see at least a face, a skull, one or more pyramids, and a bird wheeling in flight.

ESA/DLR/FU Berlin - G. Neukum

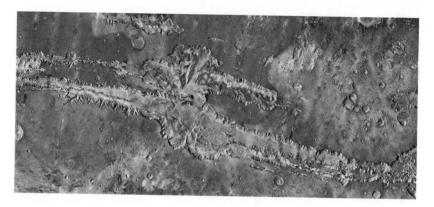

FIGURE 9.24

Valles Marineris.

NASA/JPL-Caltech/Arizona State University

distance from Los Angeles to New York City is about 3,940 km [2,450 mi]), up to 190 km (120 mi) wide, and 7 km (4⅓ mi) deep. The corresponding dimensions for the Grand Canyon are 446 km (277 mi) long, 29 km (18 mi) wide, and 1.6 km (1 mi) deep.

If, as seems likely, it was caused primarily by early tectonic plate motion, Valles Marineris is a rift valley, where two tectonic plates moved apart. Such plate motion ceased billions of years ago, as that planet cooled more quickly than the larger Earth. Different parts of the Valles Marineris have different features, some consistent with the presence, and loss, of liquid or frozen water. Many of the canyon walls have collapsed to different extents, and photographs reveal that such collapse continues, especially when impacts from space debris near or in the Valle Marineris cause the walls to vibrate. Keep in mind that as with all the worlds you could visit, such impacts still occur.

Assuming stable cliff walls can be identified, it is likely that either trails or cable cars, or both, will be installed in various places around Valles Marineris so you can travel from the top to the bottom of the canyon and back. Such a hike would be analogous to hiking down or up the Grand Canyon in the United States. We do not yet know what you will find on such an expedition.

The Poles

The poles of Mars have a variety of interesting features. The North Pole is lower in altitude than the South Pole and has a seasonal ice cap of frozen carbon dioxide (dry ice), below which lies a permanent water ice cap. As a result of being lower than the south, there is more air above the North Pole than above the South Pole. The more gas in the air, the more heat is stored in it from sunlight, so the air heats the North Pole more. This heat during the northern summer causes the entire carbon dioxide ice cap to melt.

The South Pole also has a permanent water ice cap that is not centered on the pole and an overlying carbon dioxide ice cap that is centered. The thinner air there means that this dry ice is heated less during the summer than the ice at the North Pole. As a result, some of this permanent dry ice forms a transparent layer. When summer arrives in the south, sunlight travels through this layer, heating some of the ice below it, causing some of that ice to vaporize. The resulting gas flows under the overlying dry ice, erodes the ground, and carries some of the dust from the underlying regolith until it finds a place from which to escape, often as jets of gas, into the air. The flow of dusty gas creates a spidery appearance on the surface (figure 9.25). The weathering of the ices in the south also creates a surface there that looks like Swiss cheese.

Sedimentary Layers

The Valles Marineris and other places on Mars show layering of rock on both large (figure 9.26a) and small (figure 9.26b) scales. As on Earth, there are likely several mechanisms involved in creating these features, including ash deposits from volcanoes, water flow, and wind flow. As layers build up, they compress the materials below them until they become solid—rock. Confirming the mechanisms of formation will probably require detailed examination of the various sedimentary layers.

The polar regions also have layers, which presumably result from the periodic heating and cooling of the ices combined with dust blown there in windstorms. The presence of dust, darkening the ice caps, causes them to heat up more than they would without it. As a result, the amount of

FIGURE 9.25

Spiderlike features near Mars's South Pole.

NASA/JPL-Caltech/Arizona State University

ice sublimated seasonally depends on the weather that the pole has experienced in the past year.

Volcanoes

Whereas nearly all the craters on our Moon are believed to have resulted from impacts, there are at least two dozen significant volcanoes on

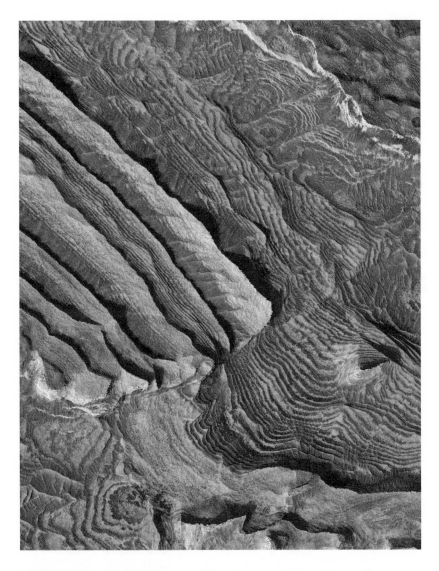

FIGURE 9.26A

Thick sedimentary layers on Mars. Each of these layers is about 10 m (33 ft) thick. The image is about 2 km (1.2 mi) across in the Arabia Terra region.

NASA/JPL-Caltech/University of Arizona

FIGURE 9.26B

Thin sedimentary layers on Mars. Each of these layers is less than 2.5 cm (1 in) thick. Taken by the rover Curiosity.

NASA/JPL-Caltech/MSSS

Mars. None of them is active today. Most notable are four: three in a row collectively called the Tharsis Montes (*montes* is Latin for mountains), plus the nearby Olympus Mons, the largest known volcano in the solar system (figure 9.27). You can see the locations of all four of these volcanoes and others in figure 9.19. Olympus Mons is about 26 km (16 mi) high and about 600 km (374 mi) wide. The largest surface volcano on Earth, Mauna Loa in Hawaii, is 10 km (6.3 mi) high (measured from the bottom of the ocean on which it rests) and 120 km (75 mi) across. Mount Everest, the highest mountain on Earth, is only 8.9 km (5.5 mi) high, so Olympus Mons ranks as the highest mountain in the solar system too. Clearly, Olympus Mons will be the destination volcano on Mars.

Like Mona Loa, Olympus Mons is a shield volcano, meaning that the lava oozed out of it rather than being ejected dramatically upward, as occurs in stratovolcanoes such as Japan's Mount Fuji. As a result of repeated slow flows, Olympus Mons does not transition smoothly into the ground around it. Rather, the edge of the volcano has a steep cliff surrounding it. This is analogous to the features called pedestal craters created by impacts from which debris is ejected, but does not make a smooth transition to the ground beyond the impact region (figure 9.28).

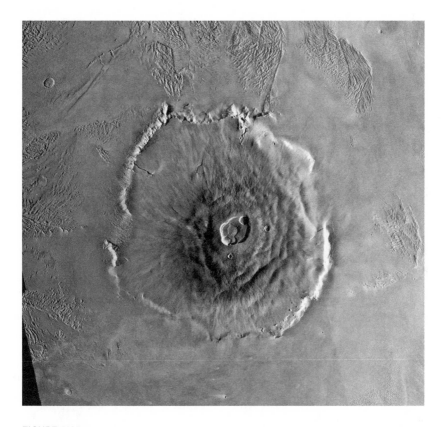

FIGURE 9.27

The volcano Olympus Mons on Mars.

NASA

As with all extinct or dormant volcanoes, Olympus Mons on Mars has a central depression at the top of its cone. Called a caldera, this depression formed when the molten rock flow stopped. Then the remaining molten rock that had been coming up from the center of the volcano solidified and shrank back down, creating the caldera (recall that except notably for water, most liquids contract when they solidify). The caldera may be accessible by a mechanized vehicle by the time you visit. Another option for extreme athletes is a trek up one of the volcanoes. Because of the length of the trek, it is likely to take several days, and oxygen, supplies, and habitats

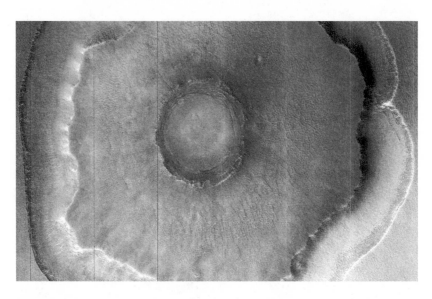

FIGURE 9.28

Pedestal crater on Mars.

NASA/JPL-Caltech/Univ. of Arizona

will have to be available along the way, as for climbing extreme mountains on Earth.

Craters

Cratering occurred on every object in our solar system except the Sun. While Mars was pummeled by many crater-forming impacts, most of those craters have been worn away by water and/or by weathering over the past four billion years. In that sense, the surface of Mars has evolved similarly to the surface of the Earth, leaving both planets with many fewer craters than our Moon, for example. Some of the remaining craters on Mars have unusual features, such as central peaks, central secondary craters, pedestal craters (see figure 9.28), a variety of different wall features, and even bodies of water ice inside them (figure 9.29). Keep in mind that the cratering of Mars (and our Moon) continues. Visiting a new crater on any world can be interesting, especially on Mars if water from inside the planet is leaking out through it, forming ice, when you arrive.

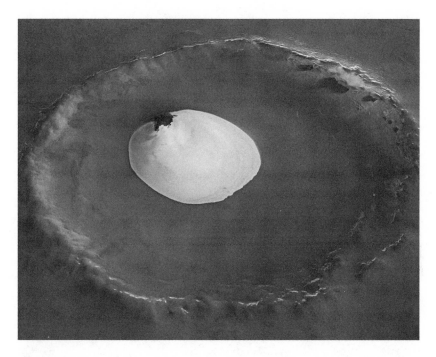

FIGURE 9.29

An impact crater with water ice.

ESA/DLR/FU Berlin - G. Neukum

Dry Riverbeds

The geological evidence for the historical presence of liquid water on Mars is overwhelming, as mentioned earlier in this chapter. One piece of that evidence is the existence of winding valleys (figure 9.30a). On Earth, such valleys are only created by liquid water flowing very gently downhill (figure 9.30b). Although rilles on the Moon created by lava flow look similar, most geologists believe that, like such features on Earth, the ones on Mars were created by the flow of liquid water. Since there is no liquid water flowing long distances on Mars today, all the winding valleys there are dry. You will be able to visit and explore one or more of these valleys. The geologist accompanying you will show you a variety of its interesting features (details of which we presently don't know).

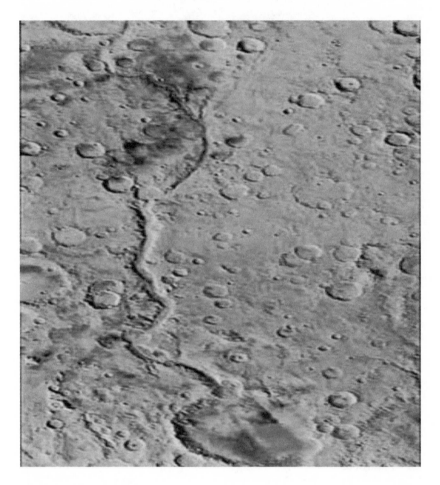

FIGURE 9.30A

(a) A dry riverbed on Mars.

NASA

Shorelines

The shorelines on Mars provide insights into the early history of that world. The shorelines of oceans and lakes on Earth are rich in debris pushed up there by wind-driven waves. As noted earlier, there is substantial evidence that the northern region of Mars was a vast ocean and that lakes dotted the early landscape. If that is confirmed, then the shorelines of these long-gone bodies of water may also hold geological and possibly biological treasures.

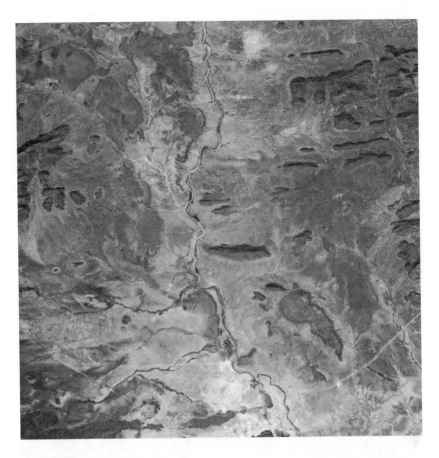

FIGURE 9.30B

(b) A flowing river on Earth.

NASA's Earth Observatory

These may include rocks rounded off by wave activity, rocks chemically changed by emersion in water, and perhaps shells of early Martian life.

Common Surface Features

A variety of other interesting features have been found on the surface of Mars that are worth seeing and, in the case of hematite, collecting. Hematite is a form of iron oxide that precipitates out of iron-rich water here on Earth, forming a crystal structure often with a hue of gray, ranging from black to silver, or rusty red or brown in color. Hematite also forms due to volcanic activity and without the presence of water. It is used, among other things, for making jewelry. Hematite on Mars appears to have formed from iron dissolved in water, although other mechanisms are still under consideration. It has been found in several sites on the planet. Most of the hematite on Mars appears as tiny spheres, called spherules (figure 9.31), typically a quarter of an inch or smaller in diameter.

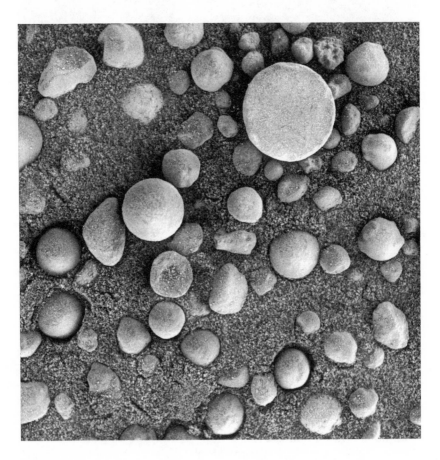

FIGURE 9.31A

Close-up of hematite spherules on Mars, sometimes called "blueberries" because of their dark gray/blue color. Image taken by the rover Opportunity.

NASA

FIGURE 9.31B

Hematite spherules among the dunes in Endurance crater. Image taken by the rover
Opportunity.

NASA

Millions of these spherules have been observed on Mars. A gallon
(3–4 l) of them would be a nice thing to bring back to Earth.

Dunes

As noted above, varying amounts of regolith dust and larger pieces of surface debris, analogous in size to what we call sand, cover Mars. As a result,
sand dunes formed there. As with everything else about that planet, there
is a lot about the surface layers of Mars that we don't yet know, including
their chemical composition over most of the planet. One thing we do know
is that the winds configure the surface layers into a variety of fascinating
patterns (e.g., figures 9.31b; 9.32a, b, and c). Visiting them will be among
the highlights of a trip to Mars.

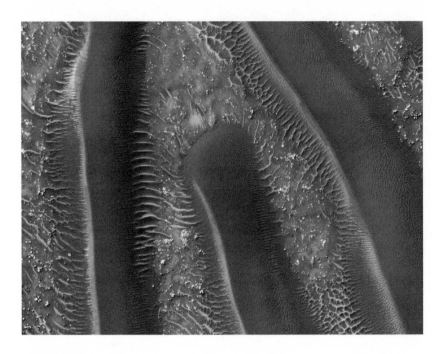

FIGURE 9.32A

Dunes on Mars. Winds blowing from the right in this figure have created the remark-able dunes shown here.

NASA

Glaciers and Brain Terrain

Mars has water ice deposits analogous to glaciers on Earth. Some of these features show fascinating structure called brain terrain (figure 9.33). Their origin and evolution are still being researched, but viewing them directly should be a great experience.

Dust Devil Tracks

We have all heard of or seen the incredibly damaging effects of tornadoes (also called cyclones), which have killed people, ripped through houses, and even leveled entire towns. Tornadoes are rapidly swirling columns of air that are formed in storms, where the differences in air pressure and

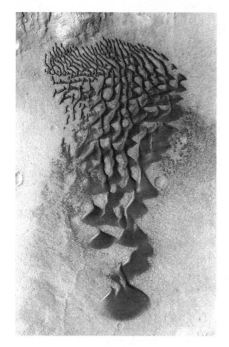

FIGURE 9.32B

Example of dunes created by winds blowing down from the top of this image.

NASA

temperature can force the air to flow in these patterns. There are much less powerful swirling flows of air here on Earth called dust devils. As they flow across the surface, dust from the ground below them is sucked up. As you can see from the two people running toward one in figure 9.34a, these are much less dangerous events than tornadoes.

Many dust devils have been observed on Mars. They can be located both visually, as shown in figure 9.34b, or using lidar (light detection and ranging) equipment, similar to radar. Because dust devils are in contact with the ground, they leave distinct patterns on the regolith (figure 9.34c). While the air pressure in dust devils on Mars is relatively low, there are possible dangers associated with them, including dust getting into sensitive equipment, chemical reactions between the dust and human equipment, static cling, and potentially toxic chemicals that could be harmful if inhaled (such as when you remove your space suit). This latter point has been proposed but not yet verified. Dust devils will be fascinating to see, but until we learn otherwise, it would be wise not to drive or run into one on Mars.

FIGURE 9.32C

Sand dunes with frost (white) and dark sand mostly overlain with lighter-colored sand. Some of the darker sand left streaks as it slid downhill, creating the impression of trees on Mars.

NASA/JPL/University of Arizona

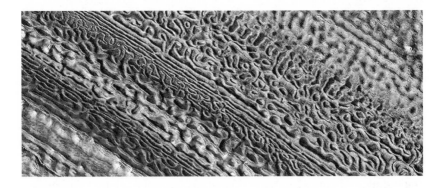

FIGURE 9.33

Brain terrain on Mars. Folds of ice that wrap around a small hill.

NASA/JPL/University of Arizona

FIGURE 9.34A

Dust devil on the Arizona desert.

NASA/University of Michigan

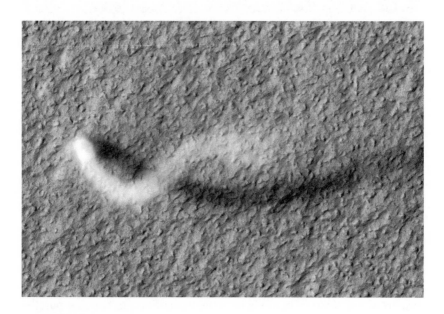

FIGURE 9.34B

Dust devil on Mars, seen from above.

NASA

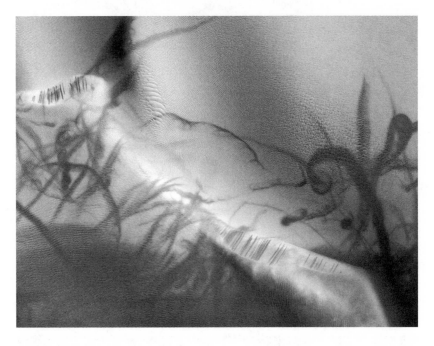

FIGURE 9.34C

Remnant tracks of dust devils on Mars.

NASA

Visiting lands covered with dust devil tracks will also be interesting, perhaps as much for the aesthetic visual impact as anything else. Indeed, they will be particularly impressive if you can see them from nearby hilltops or from an aircraft, when that technology is available for Mars. There is no known reason you can't walk over the remnants of the dust devils—they have no historic or scientific value worth preserving.

Slope Streaks

Somewhat similar in appearance to dust devil tracks, but of different origin, are slope streaks (figure 9.35). Often occurring in roughly parallel groups, these features are found on steep surfaces, such as the walls of craters or cliffs. While their origins are not yet known, they appear to be small-scale avalanches of surface debris.

FIGURE 9.35

Slope streaks running down a crater on Mars.

NASA/JPL/University of Arizona

Active Canyon Walls/Gullies

You may see that many of the craters and canyon walls on Mars have gullies and other features, some of which may have been formed by liquid water or ice flowing downhill (e.g., figure 9.36a). Whether these come from the liquid water discovered in 2015 to be flowing on Mars, past water flow, when ice periodically formed and melted, or a non-water source is not yet known. Furthermore, many craters and canyon walls are sliding downward even today (figure 9.36b), as a result of vibrations of the planet probably created by nearby impacts. When you visit the tops of canyons or craters, be very careful, as they are likely to be quite unstable. Even your walking to the edge may cause small landslides that could carry you down with them.

Your travel company may be able to arrange for you to actually initiate a landslide from a distance so that neither you nor anyone else will be in danger. This could be done, for example, with explosives. Besides the experience of seeing large quantities of debris sliding down the slope, there may be a good scientific reason for such activity. While there is no stable liquid water on the surface of Mars, there may be some in reservoirs not far below the surface. If a meteor strikes the surface above or near such a reservoir, some of the water will leak out and create gullies and other features in the crater. However, in due course, the water will drain out

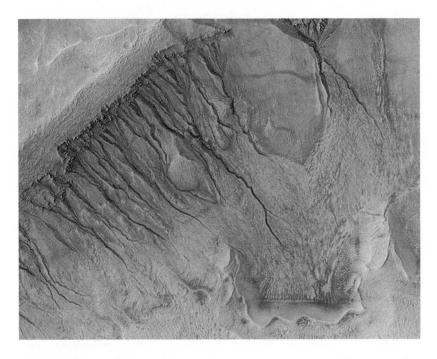

FIGURE 9.36A

Gullies and other features on the wall of crater Newton. They were probably formed by blocks of dry ice sliding down the slopes.

Malin Space Science Systems, MGS, JPL, NASA

or its flow will carry regolith that will plug the holes or cracks. To study that water firsthand, chipping away at the edges of craters (and canyons) could give scientists access to it and, if it exists, to the life in it. Another plausible explanation of some of the gullies is that they were formed by blocks of dry ice, frozen carbon dioxide, flowing downhill. Such ice on Earth sublimates (turns from solid directly to gas), creating a cushion under the block that enables it to float/slide down the slope. Which, if either, explanation is correct remains to be seen.

FIGURE 9.36B

Erosion of canyon wall in Valles Marineris.

NASA/JPL/USGS

Chaos Terrain

There are some extremely rough surface areas on Mars called chaos terrains. Several worlds in our solar system have regions where a variety of geological and astronomical events came together to create chaotic surface features unlike anything we have on Earth. For example, Mercury was hit by an impact so powerful that it sent waves (similar to sound waves) all the way through the planet. Coming out the other side, these waves caused the land there to become all jumbled (figure 9.37a). Mars has a variety of chaotic terrains with different features. While their origins are not known for sure, they are likely due to water activity, such as melting, flooding,

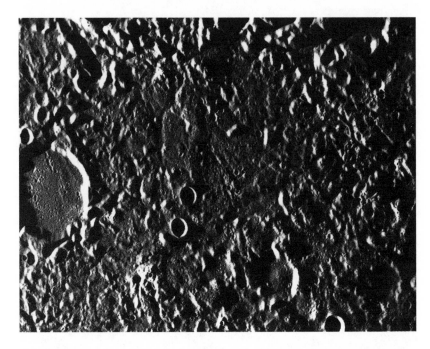

FIGURE 9.37A

Mercury's chaotic terrain on the opposite side of the planet from the Caloris basin impact.

NASA

freezing, and moving, among other things. An example of such terrain that you could visit is shown in figure 9.37b.

Sky Color/Sunrise/Sunset

Surface features aren't the only things on Mars that differ from what we see here on Earth. Even the sky there is different. While our daylight sky is a light blue, the color of the sky on Mars is normally a rusty orange—yellow, sometimes associated with the color of butterscotch. Our sky is blue because the gases in our air preferentially scatter shorter wavelengths (violet, blue, and green) compared to longer wavelengths (yellow, orange, and red). The especially intense blue light from the Sun, scattered by Earth's

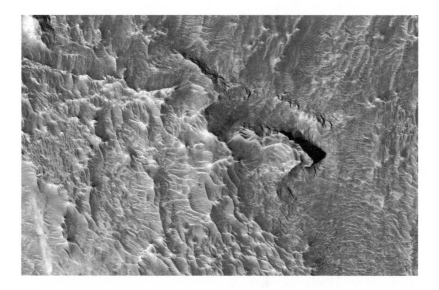

FIGURE 9.37B

Chaotic terrain on Mars. This image, taken from the eastern region of the Valles Marineris, is about 370 km (230 mi) wide.

NASA/JPL/University of Arizona

atmosphere, causes the blue sky. Lord Rayleigh (1842–1919) first worked out the details of this scattering in 1871, so it is called Rayleigh scattering. The dust particles in the atmosphere of Mars are so large, compared to the wavelengths of visible light, that another type of scattering creates the normal sky color there. It is called Mie scattering, after the German physicist Gustav Mie (pronounced "me," 1869–1957), who worked out the details in 1910. The bottom line is that the dust in Mars's atmosphere primarily scatters longer wavelengths of visible light toward the surface of the planet, giving it its rust-colored sky. When the dust level in the air is particularly low, Rayleigh scattering takes over and the sky has a bluer hue.

Interestingly, at sunrise and sunset on Mars the sky around the Sun scatters primarily blue light toward the planet, so sunrise and sunset are often blue there, even on dusty days. In summary, the colors in the sky of Mars are pretty much the reverse of what we see on Earth—our vivid orange-yellow sunrises and sunsets and our normally blue daytime sky.

Winds and Storms

Although the air pressure on Mars is only about 0.6 percent as great as the air we breathe, it has the capacity to create winds strong enough to blow a lot of the surface dust into the air. Under the right (or wrong, depending on your perspective) conditions, that wind can blow up enough dust to fill the entire atmosphere, as occurred, among other recent times, in 2001 (figure 9.38). Ten such planet-obscuring storms have been observed since 1877. They last for about half an Earth year. Many more local dust storms have been observed in various places around the planet. These latter storms normally last for days or weeks. If you encounter either a local or a global dust storm on Mars, you can expect to get a good coating of dust all over everything, on top of the static electricity-induced dust. However, as noted in chapter 2, the air density is so low that even the strongest winds there will not blow you over.

Mars • Global Dust Storm

June 26, 2001   September 4, 2001
Hubble Space Telescope • WFPC2
NASA, J. Bell (Cornell), M. Wolff (SSI), and the Hubble Heritage Team (STScI/AURA) • STScI-PRC01-31

FIGURE 9.38

Images of Mars without (left) and with a global dust storm in progress.

NASA

FIGURE 9.39

View of Mars's surface from the rover Opportunity.

NASA

Rovers and Their Trails

Just as the *Apollo* landing and other historical sites on the Moon are likely to be off limits to visitors, the rovers on Mars will also likely be preserved in situ. Including those craft that crashed onto the planet, you will have options to visit some of the landing sites (figure 9.39). Since some of the landers also had rovers, you will be able to travel from the lander to its rover.

Permanent Habitats

If all goes as well as possible, permanent colonies will be established on Mars this century. The amount of radiation Mars receives will probably lead to most of the buildings being underground (more on this shortly). Therefore, if you visit these locations, don't expect to drive through towering cities or even bustling towns with stores on ground-level streets. In underground habitats, the lighting will be cycled in brightness with the Martian day. As discussed in chapter 7, this cycling is essential for humans to function, as we are driven by biological clocks that evolved to respond to the Earth's daily 24-hour cycle of light and dark. The good news is that Mars's day-night cycle is just over 24 hours 37 minutes long. Our biological clocks can function in that time cycle, even though it is slightly longer than we are used to.

Unknown but Plausible Features

Astronomers and astrogeologists have just scratched the Martian surface. It is likely that there are a myriad of interesting things to see on or in Mars about which we don't yet know. These might include caves with a variety of features, possibly including stalactites and stalagmites; underground liquid water reservoirs; veins of minerals both on the surface and underground; and life or its remains. Stay tuned!

Part IV

HOME! SWEET? HOME?

10

EMIGRATING TO MARS OR RETURNING TO EARTH

———

EMIGRATING TO MARS

Colonizing Mars will create a new society of humans that is likely to identify itself as different from humans on Earth. As we will see in this chapter, establishing colonies on Mars will be the hardest, most expensive, most dangerous, and most transformative emigration experience in human history. Every aspect of human society will have to be modified or reinvented, including agriculture, water collection and purification, mining, manufacturing, construction, transportation, communication, medicine, reproduction, social activities, cultures, religions, education, economy, emergency responses, recreation, policing, alcohol production, and protection from radiation, to name a few. Clearly I cannot discuss any of these things in depth here—myriad books will be written on all of these matters in the years to come. Rather, I will present some main features of these elements of life on Mars.

Many people are eager to colonize Mars. Hundreds of thousands have already signed up to emigrate there. While many of them *want to go* there, many others *want to get away* from their lives here on Earth. It would be very interesting to hear these people with different motivations, often from different cultures, religions, and economic classes, discuss their reasons for emigrating.

The first tier of needs that humans will have for survival in a long-term colony on Mars includes access to safe liquid water, breathable air, food, and protection from radiation from space. However, none of these things is available on the surface of the Red Planet today, nor will any be in the foreseeable future.

Water

We have seen that water exists on the surface of Mars in the form of ice, most of which is located at the poles. Because of the low temperatures and extremely low air pressure, liquid water on the surface will either immediately freeze or immediately evaporate—recall the cup of water in the hypobaric chamber described in chapter 4. Nevertheless, there is strong evidence that liquid water exists near the surface in a variety of places. It is likely that underground liquid water on Mars will be accessible by drilling wells, just as we do in making many places on Earth inhabitable.[1]

As noted in chapter 9, I have assumed so far that the water on Mars is either lifeless or benign to humans. However, if there is life there and it is dangerous, that will change everything. First, is it possible to protect humans and other Earth-based life that goes to Mars from that Martian life? If so, that will have to be done before emigration can begin. If not, what are the ethical issues of destroying life, even simple life, on another world?

Where to locate colonies on Mars? The surface temperature varies with latitude, just as occurs on Earth. The temperature on Mars near the equator ranges from about 70°F (21°C) to about −200°F (−130°C), with an average noontime temperature of about 20°F (−7°C). Mars's rotation axis is tilted at a similar angle to the plane of its orbit around the Sun as is the Earth. This means that Mars goes through the same cycle of seasons we do. However, each season is nearly twice as long as here because the year on Mars is nearly twice as long as the year on Earth (687 days compared to 365 days).

In the best of all possible situations, colonies on Mars would therefore be constructed near the equator, where it is warmest. The lower down the habitats are built, the more air there will be above them. This air will help protect the inhabitants a little bit more from radiation from space than would be possible at higher elevations. Therefore, it is likely that the first habitats will be built in the Valle Marineris, the bottom of which is about 6.5 km (4 mi) below the rest of the planet's surface.

Breathable Air

Recall that air pressure on Mars is about 0.007 times as great as the air we breathe. Furthermore, this much, much thinner air is composed primarily of carbon dioxide, with small amounts of nitrogen and the inert element argon. Its lack of oxygen makes the present atmosphere of Mars of no direct use to the chemistry of our bodies.

Therefore, all human habitats will be sealed environments so that the breathable air is not vented out and thereby wasted. However, bringing sufficient quantities of breathable air to Mars from Earth for large numbers of emigrants would be a logistical nightmare. Air for the early settlers will have to be manufactured on the Red Planet. To make a breathable gas for habitats and air tanks, the thin, existing carbon dioxide, CO_2, and nitrogen, N_2, will be compressed, separated, and stored as liquid in tanks. Then the carbon dioxide molecules will be split into breathable oxygen molecules, O_2, and pure carbon. All this will be done using energy from the Sun. The oxygen molecules will be stored in tanks as liquid oxygen and the carbon stored in bulk form (I will discuss its use later). Finally, suitable combinations of oxygen and nitrogen molecules will be combined to create the same breathable atmosphere that we have on Earth, minus a few trace elements that are not essential for life.

As you likely know, carbon dioxide is also a by-product of respiration: our blood absorbs oxygen from the air and delivers it to all the cells in our bodies, which use it to generate energy that allows them to survive. Our blood then delivers the carbon dioxide produced in our cells to our lungs, where it is removed from the blood and expelled every time we exhale. Carbon dioxide is not intrinsically poisonous to us, as is carbon monoxide, for example. Nevertheless, most of the carbon dioxide exhaled into a Martian habitat's atmosphere must be removed[2] because high concentrations in the air prevent the carbon dioxide in our blood from leaving our bodies. Unless that carbon dioxide is removed, fresh oxygen cannot be taken in and humans will suffocate. The same applies to any closed habitat on Earth, such as a submarine.

Protection from Radiation

As noted above, inhabitants of long-term habitats on Mars (and visitors there) must be protected from radiation striking the planet from space. Recall from chapter 2 that Mars's thin atmosphere provides much less protection from this radiation than does our atmosphere, in which the ozone and carbon dioxide absorb much of the UV-B and UV-C radiations before they get to the ground, thereby protecting us from their lethal effects. Likewise, our thick atmosphere absorbs the X-rays and gamma rays from space. Because Mars has very little ozone, the low density of its carbon dioxide-rich atmosphere allows a lot of all this radiation to reach the surface.

Besides lethal electromagnetic radiation, Mars's thin atmosphere allows cosmic rays to reach the surface without spreading their energy as much as is done in our atmosphere (see chapter 7). These high-energy particles cause serious illnesses, as they smash apart organic molecules in the bodies they enter.

The combination of electromagnetic and particle radiation on the surface of Mars would therefore require that the permanent colonies be protected much more than is necessary for habitats on Earth. While it is technologically feasible to build surface buildings with walls lined with insulating water or manufactured shielding, that would be extremely expensive compared to building habitats underground. The regolith and rocks above the habitats would absorb electromagnetic and particle energy from space, keeping human and other life from Earth safe from it. Calculations suggest that at least 4–5 m (15 ft) of regolith and rocks would be necessary to create as much shielding as our atmosphere. Early emigrants to Mars will be glorified cave or cliff dwellers.

Agriculture

Along with providing water and air, feeding colonists on Mars is obviously a top priority. Shipping all the food necessary for the growing number of emigrants there is likely to be prohibitively expensive. Fortunately, figuring out how to make plant life grow in indoor habitats on Mars would lead to the production of many of the essential ingredients needed for survival. A robust agriculture could supply oxygen, food, oil, wood, medicines, plastics, rubber, and paper products, among other things.

However, like animals, plants are extremely unlikely to be able to survive on the surface of that planet. They will be destroyed by the radiation from space; frozen to death by the low temperatures that occur each night; and dehydrated by the lack of liquid water on the surface (liquid water pumped to them would almost immediately freeze). Therefore, on

Mars, plant life will have to be grown in sealed greenhouses that allow in from the sky or from artificial sources suitable amounts of the radiations needed to grow, while blocking the more hazardous ones. Water will be pumped into these agriculture environments just as it often is here on Earth. The carbon dioxide the plants require for photosynthesis will come directly from Mars's atmosphere and from the air exhaled by humans and other animals brought there. The minerals that the plants need in their soil to grow will have to be mined from Mars and mixed into the regolith in which they are planted.

The good news is that besides creating edible products, the plants that will be grown there will give off breathable oxygen as a waste product, just as they as do on Earth. Sufficient plant life there could take over producing the oxygen, which would be a great savings in money and energy compared to manufacturing the oxygen, as described above. Growing consumable food and generating breathable oxygen at the same time is a win-win situation.

The benefits of fresh fruits and vegetables and a source of breathable oxygen molecules are enormous. Furthermore, the psychological benefit of having a "green" environment for colonists to visit is likely to be profound. In order for plant life grown on Mars to provide some or most of the food necessary to sustain the colonists, the greenhouses will have to cover hundreds, and eventually thousands, of acres, like farms on Earth.

Another intriguing issue is whether it will be possible to raise animals as food stock (and pets). Otherwise, emigrants to Mars will need to be vegetarian. This issue is somewhat analogous to the issue of raising children on Mars—will mammalian bodies be able to grow and function in the low Martian gravity?

Secondary Habitat Requirements

While providing breathable air, drinkable water, and radiation protection are the primary issues for the early inhabitants of Mars, all habitats there will also need the same kinds of physical infrastructure that towns and cities have on Earth. These include electric power generation and distribution, plumbing for water and sewerage, sewerage processing, communications and transportation systems, climate control (temperature, humidity, and air purification), and emergency services (ambulances, fire vehicles, and police), among other things.

Since there are no known biologically derived resources on Mars such as oil or natural gas, the options for many of these secondary elements of a functioning society on Mars will be much more limited than on Earth. Whereas we can have transportation and long-haul vehicles run on oil-derived compounds such as gasoline, diesel, and jet fuel here on Earth, virtually all transportation on Mars will be powered by electricity (combustion engines require the presence of oxygen molecules, a very valuable commodity there). Likewise, we can use natural gas, propane, or wood to heat stoves here on Earth, but all stoves and other heating elements on Mars will be powered by electricity. The list goes on.

While electrically powered machinery is not a problem in creating a new society, providing enough electricity becomes a very big issue very quickly in light of the lack of organic fuels, such as those derived from oil and wood. The two most convenient power sources on Mars will be solar and nuclear. My best guess is that solar power will be the primary source for many years. Keep in mind that the increased high-energy radiation bathing Mars is likely to damage the solar panels designed for use on Earth, so we will need to develop more robust models.

Furthermore, without natural reservoirs of oil or natural gas, companies and countries on Mars will be unable to easily manufacture things such as plastics, natural or synthetic rubber, lubricants, asphalt, fabrics, cosmetics, and parts of virtually every device and machine you can think of. The good news is that many of these materials can be manufactured using the carbon removed from the atmosphere, along with plant by-products and, hopefully, mineable minerals on Mars.

It is important to note that we know very, very little about what is inside Mars. We don't know if all the minerals necessary to grow plant life exist. We also do not yet know how many different elements, such as iron, nickel, copper, aluminum, and sulfur, to name a few, are close enough to the surface of Mars to be mined in the near future. Without these "natural resources," manufacturing of heavy industry products, such as habitats, vehicles, tools, bridges, and machines, will also not be possible in the near future. If enough of the necessary elements are not directly available from Mars itself, many can eventually be harvested from asteroids and brought there to be used in manufacturing, but that would be a project for the distant future.

Needs of Early Colonists

As discussed above, virtually all the physical needs of the very first colonists on Mars will have to be provided by shipments from Earth. This is a logistical challenge of epic proportions. Ideally, the pieces of early habitats will be manufactured on Earth and assembled on Mars. However, this means that all the pieces must arrive there safely. Landing sufficiently gently on Mars so as not to disturb the myriad electronic, plumbing, and shielding elements built into the habitat components will be extremely challenging. Indeed, it is unlikely that all the fundamental building blocks of the habitats will arrive intact on the first try.

Terraforming Mars

Fortunately, there is not yet known to be any fundamental roadblock to this process of colonization. However, the dream of terraforming Mars, that is, creating a breathable atmosphere and copious liquid surface water, is just that: a dream. The planet does not have enough mass, and hence enough gravity, to maintain a breathable atmosphere for humans or have bodies of liquid water on the surface.

Once a molecule of water or oxygen gets into the air, it is heated by sunlight. Many of these molecules are split into separate atoms by the ultraviolet radiation from the Sun. These atoms either combine with other atoms or molecules in the air or move so quickly that they literally fly into space, never to return. Earth's mass is great enough, and hence its gravitational attraction is strong enough, to prevent most of the gases in our atmosphere from drifting into space (although some do), while the mass and gravitational attraction of Mars are not enough to hold down the oxygen and water. These factors take terraforming Mars outside the realm of science.

Social and Mental Health Needs of Colonists

Survival is necessary, but not sufficient, for the health and well-being of humans. We need to feel we are doing useful things and we need to enjoy ourselves. The colonies on Mars will therefore have to develop infrastructures within which people can have privacy when they want it, opportunities for social interactions and productive jobs, and entertainment venues, among other things.

While the hardware of survival is being put into place, it is also important to deal with the reality that people have a wide range of political, religious, and social beliefs that will affect their relationships with their fellow settlers. This will inevitably lead to the formation of groups with similar interests. How the interactions between those groups play out in an environment where you can't just up and move to another city or country is another important issue that must be addressed sooner, rather than later.

Indeed, the social interactions between the settlers on Mars will, in large measure, determine the success of the small early colonies. Without a doubt, actions that we consider criminal, such as stealing, rape, and assault, will occur. Colonies without the means of coping with people who perpetrate crimes, and a judicial system that protects innocent people accused of wrongdoing, are going to have serious difficulties. This also suggests that Mars colonies will need to have jails and other means of restraining people who are a danger to society there.

This leads to one of the most serious considerations for the early colonies, namely dealing with the mental health issues that will arise and must be dealt with constructively before people are hurt or killed. No matter how well screened they are, some people will eventually undergo mental health crises on Mars, due to biochemical issues in human bodies or social stress-induced problems, such as isolation by fellow travelers or homesickness. *Diagnoses, medicines, restraints, and therapy will be essential for the success of the colonies.*

Another crucial factor is that the work that different people do in the colonies will have different values. In other words, the colonies will develop economies that will compete with each other and eventually will affect their value to companies and countries back on Earth. Likewise, education at all levels will be integral to success. While manual labor will be needed, it is likely that higher education will be essential so that the technologies required as the colonies evolve will develop, along with an adequate supply of doctors, lawyers, engineers, scientists, social workers, and all the other specialists that make complex societies viable.

As the population of Mars grows, people there will develop different political, social, and religious beliefs, just as occurred on Earth. How will people in different regions of the planet organize themselves in states or countries, and how will these different groups interact with one another and with different countries, companies, religions, and ethnic groups here on Earth?

Should You Emigrate to Mars?

Until studies performed in microgravity and on Mars prove otherwise, current medical knowledge indicates that there are serious constraints on who should and should not emigrate to Mars. First and foremost, people who are younger than 26 should not go. While most people stop growing taller at around age 18, the human brain does not stop developing until 25, at least. If the growing brain is exposed to the level of radiation that exists in space, newly forming cells in it are likely to be damaged at a greater rate than occurs on Earth. Until the consequences of this damage are known and determined not to put young people at risk, we should opt to protect them from the radiation.

Although we have no experimental evidence, some of the consequences of taking a child through microgravity to Mars are likely to include at least profound underdevelopment of their skeletons and internal organs. Put another way, they are likely to appear malformed in adulthood, compared to their peers who grow up entirely on Earth. This is an extrapolation from the discussion of adjusting to weightlessness in chapters 6 and 7. The lack of a diversified diet for children in space could also have profound consequences for the development of their bodies and brains.

Numerous medical conditions that space medicine researchers have already identified disqualify some people from going into space as tourists today. It is a reasonable extrapolation from there to asserting that these people should not go into space for the six-month or longer trip to Mars. Such long trips are likely to exacerbate their problems and cause an unacceptably high rate of illness and early death.

Pregnant women should not go to Mars. Traveling there from Earth, their circulation systems would be seriously affected by microgravity. This change would affect the process of providing nutrients to their developing fetuses, with presently unknown consequences. Another reason for this restriction is that the fetus would suffer extreme doses of radiation on the trip, very likely to adversely affect their development in utero.

Until there is evidence to the contrary, women are advised not to get pregnant on Mars with eggs carried in their bodies, either. The problem is that these eggs are likely to have been damaged by radiation as they traveled through space. Since radiation damages unfertilized (as well as fertilized) eggs, the question then becomes: how do emigrant women who want to have children on Mars do so safely? Keep in mind that women are born with all the eggs they will ever have, whereas men continually produce new sperm. Both eggs and sperm exposed to radiation in space could be damaged.

Nevertheless, healthy, viable eggs could be brought to Mars. First, eggs would be removed from emigrant women while still on Earth. These eggs would be preserved by freezing, as is often done today. The frozen eggs, either unfertilized or fertilized (forming an embryo) could then be encased in a protective radiation shield carried deep inside the spacecraft going to Mars. While the protection is unlikely to block 100 percent of the radiation striking it, the technology is or will soon be available to give the frozen eggs or embryos as much protection as does the Earth's atmosphere for eggs and embryos here today. The eggs or embryos would then be put into the uterus in the protected environment of a Martian habitat, in which the pregnant woman should remain (to avoid harmful doses of radiation) until the child is born.

While remaining underground, protected from radiation and cosmic rays, may suffice to prevent fetuses from being damaged, those children will grow under different gravitational conditions than humans here on Earth. Plausibly, this will not be a lethal problem, but it will definitely change their physical development, most notably in the formation of their skeletons and muscles.

Having less gravitational pull to fight, their skeletons need not be as sturdy as ours; nor will their muscles have to be as strong, as they will not have to do as much work maintaining their bodies. This includes their heart muscle. As a consequence of the different physical development that children on Mars will undergo compared to their peers on Earth, it is exceedingly likely that the Mars-grown humans would not have the capability of functioning on Earth. Their skeletons would not be strong enough to bear their weight and enable their motion here, and their hearts would not be strong enough to circulate blood in Earth's gravity as effectively as it needs to be moved. The list of other physiological differences caused by growing up in Mars's lower gravity could be extensive. The bottom line is that children who grow up on Mars could be a new species. Would we consider them aliens?

Social Constraints to Emigration

While people of virtually every society, social class, education level, religion, and sex and sexual orientation, among other classifications, may want to emigrate to Mars, there are several social reasons that people should think very carefully about making that commitment. Foremost is that it is extremely unlikely that emigrants will ever be able to come back to Earth. While people have emigrated from their countries of origin since time immemorial, they have all had the physical option of going home one way or another. Even if they may not have had the political, social, emotional, or economic means to return, it would at least have been physically possible. Besides the considerable economic cost of returning home from Mars, the physical changes that people undergo in low-gravity conditions, discussed in chapters 6 and 7, would make the return after an extended stay very perilous, with a high likelihood that they would suffer debilitating, if not fatal injuries shortly afterward.

Homesickness

While the glamour of space travel will likely carry most emigrants on the journey to Mars, it is likely that at some time during that outbound trip they will experience serious homesickness, as discussed earlier. Intellectualizing the fact that homesickness comes to most emigrants is one thing, but feeling it grinding you down and taking the gilt edge off the voyage is something else. Symptoms of homesickness are listed in chapter 8.

Most people overcome these symptoms, but they can leave permanent emotional scars that can profoundly affect how you feel thereafter. Some meds can help overcome some of the symptoms of severe homesickness. Nevertheless, it can color your travel experience for years.

People Who Need You on Earth

While homesickness will affect emigrants to Mars, the impact of never seeing these travelers again may be devastating to some of the people they leave behind. What might happen if a young parent has and accepts the opportunity to emigrate to Mars? They would have to leave their children behind. Of course, parents have had to emigrate without their children over history, but the vast majority of those children are led to believe that they will see their parents again, either by eventually joining them in the new land or when the parents come back.

The expectation of reunion is then part of a child's life, if their parents have emigrated on Earth. While grown children of Mars émigrés may have the opportunity of eventually joining their parents on Mars, that would require them to leave their lives and friends behind. Whereas children of émigrés historically have left impoverished living and social conditions when joining their parents on Earth, that is unlikely to be the case for the grown children of émigrés to Mars. The children would have to leave friends, lovers, extended families, jobs, recreational activities, and more for an unknown life on a remote, relatively sterile world with little prospect of returning to our verdant planet.

FOR ALL SPACE TRAVELERS RETURNING TO EARTH

I had grown weaker and my bones were softer. I was no longer an earthling but a fully adapted, lived-there-all-my-life space-man. In the back of my mind was the nagging concern that my transition back to the planet might prove more difficult than my adaptation to space had been.

—Jerry Linenger, *Off the Planet*

Psychological/Social Readaptation

When I was in my early twenties, I went on my own to live in Europe for three years. I initially knew no one there, nor the customs of the country to which I moved (Wales). I was a graduate student in what was then arguably one of the most challenging fields of science, Einstein's general relativity. For months my whole life revolved around the calculations I was doing. It was only after I got comfortable with the business end of things that I was able to relax and enjoy a wide variety of interesting and unusual experiences, including: going down a working coal mine; driving across Europe; standing on the top of Mount Rigi, Switzerland, looking over the snow-capped Alps; listening to world-class music in Salzburg; spending a week on a mountain in Sicily; and presenting my research to Stephen Hawking and his group at Cambridge, among many other things.

And then I went home.

There were my family and friends, behaving pretty much as I had remembered them from three years earlier. But I was a truly different person than the me who had left the States those three long years before. I had seen and done things that had changed me in ways that words could only begin to describe. My view of the world had changed. For example, standing at the top of the Alps, I found myself thinking about large-scale features of the Earth that I had never considered before. How had the mountains formed? How were they related to each other and to the rest of the world? They brought a sense of beauty and order to a world filled with human turmoil. The change in my perspective, however, pales in comparison to how differently you will likely feel when you come home from space, even after a very short visit.

By the end of your time in space, you will have been away from Earth for hours, weeks, months, or perhaps years. You will have seen and done things that few humans have ever seen or done before. Perhaps as a child you dreamed of flying like Superman, and now, in space, you will have done it. You will have taken thousands of photographs. You may have touched another world, feeling its textures, smelling its smells, and seeing its sights. If so, you will have without a doubt compared that world with the Earth and come to realize that the two bodies are not the same.

You will then have a new perspective on our planet that will change you forever.

Every time you touch one of the souvenir samples of the world(s) that you visited, it will bring you back there in ways that will be hard to describe to your friends and family. Your trip will change much more than your views of the different places in our solar system. If all goes well, in space you will have developed new friendships and a strong sense of community with your fellow travelers, tourists and crew alike, or at least with some of them. You will have become proficient in working with some of the most complex engineering and scientific equipment of our time, which will give you a sense of competence and, well, superiority to the folks back home, who can only glimpse such high technology on the web or in sci-fi movies. If it was long enough, your trip might also have given you time to work on college courses you've always wanted to take but never had the time for. Then there were the space walks and the unbelievable feeling of floating in space, free and unlimited. It's a good thing there were others out there to drag you back inside, because you *really* may not have wanted to go back into the spacecraft.

Perhaps most significant is that you will have seen the Earth as a whole, much more complete than what I saw from the top of Mount Rigi. That view you will have had, and probably spent hours watching, changes everyone who sees it. What a spectacular feeling it is to see an entire world through the ship's windows from deep space, especially when that world is the Earth. By all reports, it is a feeling you can only get by being there. It isn't just the view . . . it's knowing that there, below, is home. Your home, your family's home, the origin of all life as we know it. Life began on that blue, brown, and white ball you saw in the window, teeming with the most complex, beautiful, unique, and meaningful things in the known universe.

Thinking about the Earth as a single entity as you viewed it in space might bring you to question how it is that people on the surface have such narrow perspectives about our planet, life, and everything. You may wonder why we can't manage to live together without causing damage and creating bad feelings toward each other. After all, we are humans; all members of the same species; all living on the same world. And you will know from your time in space how fragile and unique is that world, the Earth. There is *nothing* else like it in our corner of the universe.

Of course, your trip wasn't all a bed of roses. You made the physical adjustments to space; your face swelled up in the first few days, and during that time you experienced a constant need to urinate. You would rather not recall the flu-ish feeling, nor the vomiting that space sickness brought on. But all that passed, even the difficulty sleeping that dogged you through much of the flight. You also got through the disorienting and uncomfortable perceptual and psychological adaptations, such as learning to have a conversation with someone who appears upside down.

If you went on a long trip, there may well have been long periods (or at least they seemed long) when half of the ensemble wouldn't talk to the other half or when someone expressed disgruntlement inappropriately. You might have conquered a bout of claustrophobia with help from the psych folks. Perhaps a fellow passenger broke her ankle and suffered throughout the trip. Everyone on board suffered with her, while being thankful that the injury didn't happen to them. You might have endured a bout of depression when you got bad news from home and found that the on-board psych meds were only marginally helpful. Now imagine that the landing is only a few hours away and that the many people you love will be waiting for you.

Psychological Readaptation

Homecomings are often exciting times, when people reunite, stories and gifts are exchanged, plans are made, and lives resumed. However, not far beneath the façade of happiness and togetherness lurks a period of readjustment and testing. Based on the experiences of people who have wintered over in Antarctica, returned from submarine cruises, emerged from voluntary group confinement experiments, or come back from long times in space stations, we know a little about what you can expect as you readapt to Earth and the relationships you had before you left.

Foremost, your adventure in space is going to set you apart from people who have not gone there. You will attain a level of celebrity because of the adventure, although it won't be anything as profound as the celebrity experienced by the first few generations of astronauts. Nevertheless, people who know you have spent time "out there" will want your autograph and to take selfie pictures with you. You, on the other hand, will see everyone else as "them" for a long time, including your family and friends.

You, your family, and your friends all will have changed while you were away. People who have been separated from their primary family and friends and who have had a close relationship with another group of people for a long period of time develop a new set of interpersonal behaviors. Going back to their former "self" is often hard, if not impossible. You *aren't* the same person you were when you left, just as the people you were close to on Earth aren't the same. People continue to grow and change throughout their lives. Normally those with whom they are close are part of those changes so that everyone continually adapts to one another as life goes on. The problem is that when you return from a long time far away, you and the people you "left behind" have changed *differently*—even Skyping daily, you weren't part of each other's changes, so you don't "know" them like you did.

Readjustment needs to be your primary goal upon return to Earth. Space travel amplifies the period of readjustment and the testing of relationships caused by your absence. Even if the trip is only a matter of weeks, you will have been deeply affected and changed by the experience in space, in ways that people who haven't been there really can't understand. Space travelers report that the journey changed their views of the meaning of their own lives, the meaning of life as a concept, and their religious and political beliefs.

As former astronaut Dr. Jerry Linenger writes in his book *Off the Planet,*

> I have been a U.S. naval officer for twenty years. I understand the necessity of armed forces. But I have also seen the undivided Earth from space. When viewed from this perspective, the fighting amongst ourselves makes no sense whatsoever. Now, whenever I witness conflict in any form, I try to step back and examine the problem from a broader perspective. Understanding follows. (247)

Sailor Alvah Simon, who spent a solo winter trapped in his sailboat in the Arctic, described the poignancy of reading the famous "I will fight no more, forever" speech by Nez Perce First American Chief Joseph. "I closed the book and cried so hard I thought my heart would break, just as his had. In the darkness, in my loneliness, I had never in my life felt more closely connected to humankind." The reality is that some of the things that people experience in space change them, sometimes making it difficult, if not impossible to live with the people they were close to before they left and who held different views that have not changed.

Readjusting to or shedding relationships that you had on Earth before you left is only part of the process of readjustment when you return. Although much of the time you spend in space will be taken up by the normally mundane process of getting from one place to another, it will be punctuated by periods of extreme excitement, elation, awe, and terror at what you see and what happens around you. It is likely that you will go through a phase upon returning to Earth when you look around at the attractive, but relatively everyday things on our planet and feel a sense of disconnect between the two experiences of living in space and living on Earth. This dichotomy is likely to be confounded by Earth's incredibly lush, complex, varied, and magnificent biome compared to the sterile beauty you saw in space. You will have led two different lives in two very different environments, and you now will have to reconcile them.

The commodities available on Earth that you will rediscover are going to come as a jolt. The differences between the safe, organized, reliable, repairable, disposable, and abundant products available here and the ones you had to put up with, maintain, repair, and reuse for months or years in space will exacerbate the contrast between life on Earth and the life you led up there. On top of that will be the jolt from getting back into the regimen of paying bills, going to work, preparing dinner, filling the car with gas, and all the other mundane daily activities that you left behind when you departed for space. We know from people who have lived in Antarctica and on submarines that all these contrasts can create a sense of isolation from other people that can easily take a year or more to recover from. People coming back from Antarctica, especially, experience severe depression at greater rates than the population at large, as well as higher rates of alcoholism and suicide attempts than the average population. In *Mountains of Madness: A Scientist's Odyssey in Antarctica*, John Long writes:

> For about three months after returning to Australia I was on a bit of an emotional roller coaster, sometimes crying without reason at sad movies or emotional events, or laughing wildly at simple, stupid things. Maybe my

mind was just letting off the accumulated steam of the trip. My emotions eventually seemed to have mellowed out into the realm of normality but, to be honest, have never really been exactly the same as before the trip.

Prepare to be changed forever.

Biological Readaptation

Your physical readaptation will also commence the minute you land on Earth again. Because of the physiological changes discussed in chapters 6 and 7, your bones, circulatory system, muscles, sense of balance, posture, and sleep cycle all will have become acclimated to life in space, but they would be unsuitable for living back here in Earth's gravity. When Dr. Linenger was reunited with his family after four months of living in the *Mir* space station, his brother later told him that "he had been shocked by my appearance the first time he saw me after the flight. To him I looked thin and weak, my flesh pale. I moved unsteadily and looked like I had not slept in weeks. My handshake was a rather feeble one" (233).

Since your sense of your physical self compared to your surroundings (your proprioception) will be different when you return, you are likely to find yourself moving differently or inappropriately even when you do simple things like reaching for something on a shelf. Your immune system will also take time to recover, meaning that for a while you will be more susceptible to illness than you were before you left.

After you return from a trip of a month or more, your physical readjustment to Earth will go in stages from days (adjusting to changes in posture, such as sitting up or rising without feeling lightheaded or fainting; proprioceptive activities, like driving) to weeks (a sense of balance[3] and walking; the ability to move both eyes together) to years (restoring muscle and bone mass; sleep). You are also likely to experience back pain as your spine recompresses under the Earth's gravitational force. Dr. Linenger vividly describes how even a humble shower nearly got the best of him after his time on *Mir*: "The spray of water from the shower was like pellets bombarding my body. I felt as if I would be sent tumbling. My mind was not yet Earth-adjusted, and such a force in space would have caused a reaction—pushing me away from the stream of water. . . . For a while I braced and tried to withstand the power of the shower pellets. I eventually gave up. I resorted to taking my often dreamed-of first glorious shower back on the planet sitting on the shower floor with the water dribbling out of the showerhead" (234). He then began a physical rehabilitation program to regain bone and muscle mass. He comments on having a sense of being easily injured, of physical vulnerability. After a strenuous rehab recovery, he also noted that "re-establishing the nerve to muscle conduction paths seemed to be the most stubborn deficit . . . I did not feel fluid and natural running for almost a year after my return to Earth" (237). I must say that he looked fit and functional when I saw him years later.

You may also experience sleep disturbances back on Earth, for both physical and psychological reasons. Your body has to readapt to sleeping in a normal 1g environment that also has more profound changes in the day-night cycle rhythm of bright and dark, noisy and quiet, warmer and cooler, among other things, than you experienced in space. Dr. Linenger wrote: "Gravity now yanked me down into the mattress" (234). Some people returning from Antarctica report having sleep disturbances for two years.

After any trip off the Earth for more than a few weeks, you will be strongly advised to leave the landing craft on a stretcher. While you may feel that this lacks the dignity and triumph of a heroic return, it will help prevent you from breaking bones weakened by loss of calcium, passing out due to too little blood in your circulatory system, bumping into things because your proprioception and depth perception are not working, or falling because your muscles can't hold you up. Indeed, Dr. Linenger confides that his stern remonstrance to himself before appearing in public was, "Whatever you do, Jerry . . . don't pass out." Families of the returning space travelers will be lining the red carpet on which you will be rolled from the lander, but you may only be able to raise an arm and wave to them.

While they might look different to you after months or years, you will definitely look different to your friends and family. Assuming that your spaceship did not have artificial 1g gravity, you will almost certainly have lost a lot of weight. You will be much weaker than you were when you left Earth since your muscle mass will have decreased in space. Your posture will be very poor, as measured by Earth standards, as you will have been comfortably slouched over for almost your entire trip.

It is likely that everyone returning from a long space voyage will have a private room in which to greet family and friends and to exchange gifts. If you went to Mars or other locations away from the Earth-Moon system, you may not be able to get off the bed into which you are carried by staff, or to lift children. Of course, everyone on the ground will have been briefed about what to expect when they see you, but still in many ways it will be a struggle, if also a joy, to reestablish your life on Earth.

POWERS OF TEN

—

Astronomy is a science of extremes. As we examine various cosmic environments, we find an astonishing range of conditions—from the incredibly hot, dense centers of stars to the frigid, near-perfect vacuum of interstellar space. To describe such divergent conditions accurately, we need a wide range of both large and small numbers. Astronomers avoid such confusing terms as "a million billion billion" (1,000,000, 000,000,000,000,000,000) by using a standard shorthand system. All the cumbersome zeros that accompany such a large number are consolidated into one term consisting of 10 followed by an exponent, which is written as a superscript and called the "power of ten." The exponent merely indicates how many zeros you would need to write out the long form of the number. Thus,

$$10^0 = 1$$
$$10^1 = 10$$
$$10^2 = 100$$
$$10^3 = 1000$$
$$10^4 = 10,000$$

and so forth. Equivalently, the exponent tells you how many tens must be multiplied together to yield the desired number. For example, ten thousand can be written as 10^4 ("ten to the fourth") because $10^4 = 10 \times 10 \times 10 \times 10 = 10,000$.

In scientific notation, numbers are written as a figure between 1 and 10 multiplied by the appropriate power of 10. For example, 273,000,000 can be written as 2.73×10^8. The distance between Earth and the Sun can be written as 1.5×10^8 km. Once you get used to it, you will find this notation more convenient than writing "150,000,000 kilometers" or "one hundred and fifty million kilometers."

This powers-of-ten system can also be applied to numbers that are less than 1 by using a minus sign in front of the exponent. A negative exponent tells you that the location of the decimal point is as follows:

$$10^0 = 1.0$$
$$10^{-1} = 0.1$$
$$10^{-2} = 0.01$$
$$10^{-3} = 0.001$$
$$10^{-4} = 0.0001$$

and so forth. For example, the diameter of a hydrogen atom is approximately 1.1×10^{-8} cm. That is more convenient than saying "0.000000011 centimeters" or "11 billionths of a centimeter." Similarly, 0.000728 equals 7.28×10^{-4}.

Using the powers-of-ten shorthand, one can write large or small numbers like these compactly:

$$3{,}416{,}000 = 3.416 \times 10^6$$
$$0.000000807 = 8.07 \times 10^{-7}$$

Because powers-of-ten notation bypasses all the cumbersome zeros, a wide range of circumstances can be numerically described conveniently:

one thousand $= 10^3$
one million $= 10^6$
one billion $= 10^9$
one trillion $= 10^{12}$

and also

one thousandth $= 10^{-3} = 0.001$
one millionth $= 10^{-6} = 0.000001$
one billionth $= 10^{-9} = 0.000000001$
one trillionth $= 10^{-12} = 0.000000000001$

NOTES

---

1. SCIENCE AND THE SOLAR SYSTEM OVER EASY

1. NASA (National Aeronautics and Space Administration) uses more than 800 abbreviations in its vocabulary. They have reference lists of them to help newbies. However, with the exception of a few acronyms such as NASA, CCD (shorthand for charge-coupled devices, used in cameras), and PTSD (post-traumatic stress disorder), I will be spelling out nearly all the expressions that are commonly abbreviated.
2. There are known to be planets orbiting other stars. These are called "exoplanets" or extrasolar planets.
3. While many numbers in science can be determined to high precision, it is often more confusing than instructive to provide too many details. Therefore, I will use terms like "roughly" and "about" so that you get useful ballpark numbers that can, for example, be comfortably compared to each other or to numbers you know already.
4. Makemake's first moon was discovered as I was writing this chapter.
5. Many of the numbers in this book are most efficiently presented as powers of ten. The appendix presents the formalism for this notation.

2. BRIEF DESCRIPTIONS OF JOURNEYS THROUGH SPACE

1. This means that every second something is falling, its velocity increases by 9.8 m/sec (32 ft/sec). For example, an object falling at 9.8 meters per second (32 ft/sec) now will be falling at a speed of 19.6 m/sec (64 ft/sec) one second later.
2. One of the very first computer games, run by submitting a deck of cards containing the program to the computer, was called Lunar Lander. In the game, you used a keyboard to control the descent of a lunar lander. If you landed successfully, an astronaut climbed out of the lander and got lunch at McDonalds on the Moon. You may have that opportunity, but the issue of meals in space, discussed in chapter 6, will require McDonalds to adjust their menu.

4. TRAINING FOR SPACE TRAVEL

1. As discussed in chapter 2 concerning travel to distant objects and above regarding the centrifuge, it is possible to simulate the effect of gravity by being spun around a central axis. The problem is that if you are close to the axis of rotation, you would have to spin very fast to re-create the normal force of gravity. That rapid rotation would itself make you sick. Therefore, to make this work, you have to be in a hotel or vehicle that is sufficiently large in diameter and spinning very slowly so that you don't feel sick. This is likely to occur sometime in the future.

6. ADJUSTMENTS DURING THE FIRST FEW DAYS

1. The other major reason, of course, is that so many astronauts experience space sickness.

7. LONG-TERM PHYSICAL ADJUSTMENTS TO SPACE

1. Studies of the mechanics of bone activity show that these soft regions do not seriously decrease the strength or rigidity of the bones. Almost all the strength and structure of any solid object, like a bone or a tree, is in the outer half of the object. That is why trees can rot on the inside and still remain standing for years.
2. As a senior in college, I did a "cosmic ray lab" in which I saw the cosmic rays passing through a cloud chamber in front of me. It was eerie realizing that other cosmic rays were also passing through me.
3. This shielding is often layers of plastic that break apart when struck by particles, like the atoms in our atmosphere. It has been proposed to encase spacecraft in a shell of liquid water, which would also absorb many cosmic rays and result in cascades of secondary cosmic rays. That water could also be used for normal activities, such as drinking, and then it would be recycled into the shell.
4. http://www.nature.com/articles/srep34774.
5. As are the impact craters found on Earth.

8. GETTING ALONG IN SPACE: PSYCHOLOGICAL AND SOCIAL ASPECTS OF SPACE TRAVEL

1. I have chosen the word "ensemble" to include all the people (both crew and passengers) in your spacecraft. I will use the more common word "group" for specific combinations of people.
2. Interestingly, there is now what is called "virtual human" software that can be used for counseling on a variety of problems with excellent results. It was first developed to help

soldiers suffering from post-traumatic stress disorder (PTSD, discussed later in this chapter) cope with their illness. This software enables you to talk to a computer-generated image of a real person whose facial expressions are so realistic that you can easily forget it is a computer to which you speak. Furthermore, the database for virtual humans is so complete that you can ask the virtual counselor any of thousands of questions and get meaningful answers.

3. This number comes from: .03 mental illness events per person per year × 12 people × 3 years = 1 person.

4. I use the word "work" here in a very general sense. It need not be in your field of expertise on Earth. You might choose to get training in some new field while you are in space and then apply that education while you are out there.

5. Both drugs and alcohol are banned in space, although there are reports that some have been smuggled on board.

9. EXPERIENCES BY DESTINATION

1. Open clusters of stars are groups of between a few and perhaps a thousand young stars that form together, but then drift apart over millions of years. Our solar system is likely to have formed in an open cluster. Globular clusters contain upward of hundreds of thousands of stars that are gravitationally bound together.

2. I say "with the Earth" because contrary to intuition, the Moon does not orbit the Earth. Rather, they both orbit a common center of mass called the barycenter, located 1,712 km (1,064 mi) below the Earth's surface in a straight line between the centers of the two worlds. The same is conceptually true of any pair of orbiting bodies, such as the Earth and the Sun. Likewise, think of two dancers waltzing down a dance floor.

3. Eclipses don't occur on every full and new Moon because the Moon orbits the Earth in a plane that is at an angle of about 5° to the ecliptic. Therefore, when it is new or full, the Moon is often slightly above or below the ecliptic. At those times, the shadows of the Moon or Earth, respectively, don't strike the other body and an eclipse does not occur.

4. Accepting the principle of Occam's razor does *not* prevent scientists from exploring other explanations of anything. Scientists are frequently trying to develop better explanations for the things they study, meaning more accurate (or with fewer unproven assumptions). When scientists achieve a theory that provides a more accurate explanation of something, their theory replaces the preceding one, even if the new one is more complicated.

10. EMIGRATING TO MARS OR RETURNING TO EARTH

1. If it turns out that this projection is optimistic and the liquid water footprint on Mars is much smaller than we would need or nonexistent, then there will be two options for acquiring water. One is mining ice and setting up colonies a practical distance from this

ice (closer to the poles than to the warmer and more desirable equator). The other is to capture comets and send them to Mars or to mine their water in space.

2. Remaining CO_2 can be used, for example, to enable people to grow plants for recreation in their habitats.

3. Some astronauts and cosmonauts report that their balance is not good for years after returning to Earth.

BIBLIOGRAPHY

To help you find relevant material more rapidly, I have separated the references by the major topics to which they relate. Some of the books and websites cited here provide information about the topics in several chapters. These are listed under "General Information" or under a specific body, such as Mars or the Moon. Some references are cited in more than one section.

GENERAL INFORMATION

Asashima, M. and G. M. Malacinski, eds. *Fundamentals of Space Biology*. Tokyo: Japan Scientific Societies Press and Springer-Verlag, 1990.

Ball, J. R. and C. H. Evans Jr., eds. *Safe Passage: Astronaut Care for Exploration Missions*. Washington, DC: National Academy Press, 2001.

Cheston, T. S. and D. L. Winter, eds. *Human Factors of Outer Space Production*. AAAS Selected Symposium 50. Boulder, CO: Westview Press, 1980.

Clay, R. and B. Dawson. *Cosmic Bullets*. Reading, MA: Helix Books/Addison-Wesley, 1997.

Collins, P. and K. Yonemoto. "Legal and Regulatory Issues for Passenger Space Travel." http://www.spacefuture.com/archive/legal_and_regulatory_issues_for_passenger_space_travel.shtml.

Comins, N. F. and W. J. Kaufmann III. *Discovering the Universe*. 10th ed. New York: W. H. Freeman, 2014.

Drury, S. *Stepping Stones: The Making of Our Home World*. Oxford: Oxford University Press, 1999.

Festou, M. C., H. U. Keller, and H. A. Weaver, eds. *Comets II*. Tucson: University of Arizona Press in collaboration with the Lunar and Planetary Institute, 2004.

Freeman, M. *Challenges of Human Space Exploration*. Chichester, UK: Praxis, 2000.

Harris, P. R. *Living and Working in Space.* 2nd ed. Chichester, UK: Praxis and John Wiley, 1996.

Huntoon, C.L.S., V. V. Antipov, and A. I. Grigoriev, eds. *Space Biology and Medicine Volume 3: Humans in Spaceflight.* Reston, VA: AIAA Press, 1997.

ISS Benefits for Humanity. http://www.nasa.gov/mission_pages/station/research/benefits/index.html.

Landisa, R. R. et al. "Piloted Operations at a Near-Earth Object (NEO)." *Acta Astronautica* 65 (2009): 1689–1697.

Larson, W. J. and L. K. Pranke. *Human Spaceflight: Mission Analysis and Design.* New York: McGraw-Hill, 1999.

Lewis, J. S. *Physics and Chemistry of the Solar System.* Rev. ed. San Diego: Academic Press, 1997.

"Man-Systems Integration Standards, Revision B." http://msis.jsc.nasa.gov/.

McNamara, B. *Into the Final Frontier: The Human Exploration of Space.* Fort Worth: Harcourt, 2001.

"Mission Preparation and Prelaunch Operations." http://science.ksc.nasa.gov/shuttle/technology/sts-newsref/stsover-prep.html.

"New Regulations Govern Private Human Space Flight Requirements for Crew and Space Flight Participants, FAA." http://www.faa.gov/about/office_org/headquarters_offices/ast/human_space_flight_reqs/.

Ordyna, P. "Insuring Human Space Flight: An Underwriter's Dilemma." *Journal of Space Law* 36 (2010): 231–251.

"Phantoms from the Sand: Tracking Dust Devils Across Earth and Mars." http://www.nasa.gov/vision/universe/solarsystem/2005_dust_devil.html.

Press, F. and R. Siever. *Understanding Earth.* 3rd ed. New York: W. H. Freeman, 2001.

Shayler, D. J. *Disasters and Accidents in Manned Spaceflight.* Chichester, UK: Praxis, 2000.

Siegel, K. "Forging Into the Final Frontier." http://www.riskandinsurance.com/forging-final-frontier/.

Space Data. 5th ed. Redondo Beach, CA: Northrup Grumman Space Technology, 2003.

"Space Flight Participants in Government Space Programs, NASA." https://www.google.com/url?sa=t&rct=j&q=&esrc=s&source=web&cd=3&ved=0CCsQFjACahUKEwjS3rvkr_zHAhWHqh4KHU2JDkQ&url=http%3A%2F%2Fwww.dsls.usra.edu%2Feducation%2Fgrandrounds%2Farchive%2F2008%2F20080826%2Fdavis.pdf&usg=AFQjCNFhv88Vucm6uqPNJlWuhI4a7awsiw.

"Space Stations." http://www.scienceclarified.com/scitech/Space-Stations/index.html.

Tyson, N. D. *Space Chronicles.* New York: Norton, 2013.

"The U.S. Commercial Suborbital Industry: A Space Renaissance in the Making." https://www.google.com/url?sa=t&rct=j&q=&esrc=s&source=web&cd=1&cad=rja&uact=8&ved=0CB4QFjAAahUKEwjEz7O6sPzHAhWGXB4KHYucBbo&url=https%3A%2F%2Fwww.faa.gov%2Fabout%2Foffice_org%2Fheadquarters_offices%2Fast%2Fmedia%2F111460.pdf&usg=AFQjCNGC3PTjm7Qu7XL9FywHYLG-ggkDAA.

MARS

"Common Surface Features of Mars." https://en.wikipedia.org/wiki/Common_surface _features_of_Mars#Mantle.

"Fretted Terrains and Ground Deformation." http://www.jpl.nasa.gov/spaceimages/ details.php?id=PIA17571.

Hoffman, S. J. and D. L. Kaplan, eds. "Human Exploration of Mars: The Reference Mission of the NASA Mars Exploration Team." 1997. http://www.nss.org/settlement/ mars/1997-NASA-HumanExplorationOfMarsReferenceMission.pdf.

Hopkins, J. B. and W. D. Pratt. *Comparison of Deimos and Phobos as Destinations for Human Exploration and Identification of Preferred Landing Sites*. AIAA SPACE 2011–7140 Conference & Exposition, September 27–29, 2011, Long Beach, California.

Kieffer, H. H., B. M. Jakosky, C. W. Snyder, and M. S. Matthews, eds. *Mars*. Tucson: University of Arizona Press, 1992.

Kress, A. M. and J. W. Head. "Ring-Mold Craters in Lineated Valley Fill and Lobate Debris Aprons on Mars: Evidence for Subsurface Glacial Ice." *Geophysical Research Letters* 35 (2008): L23206. doi:10.1029/2008GL035501. www.planetary.brown.edu/ pdfs/3604.pdf.

Levy, J., J. W. Head, and D. R. Marchant. "Concentric Crater Fill in the Northern Mid-Latitudes of Mars: Formation Processes and Relationships to Similar Landforms of Glacial Origin." *Icarus* 209 (2010): 390–404. www.planetary.brown.edu/pdfs/3798 .pdf.

"Mars as Art." http://mars.nasa.gov/multimedia/marsasart/.

"Martian Glaciers and Brain Terrain." http://hirise.lpl.arizona.edu/ESP_033165_2195.

"A New Way to Reach Mars Safely, Anytime, and on the Cheap." http://www.scientific american.com/article/a-new-way-to-reach-mars-safely-anytime-and-on-the-cheap/.

Oberg, J. "Red Planet Blues." *Popular Science* 263, no. 1 (July 2003).

Petranek, S. M. *How We'll Live on Mars*. New York: Simon & Schuster/TED, 2015.

Roach, M. *Packing for Mars*. New York: Norton, 2011.

Safe on Mars: Precursor Measurements Necessary to Support Human Operations on the Martian Surface. Washington, DC: National Academy of Sciences.

Shayler, D. J., A. Salmon, and M. D. Shayler. *Marswalk One*. Chichester, UK: Praxis, 2005.

"Slope Streaks on Mars: IAG Planetary Geomorphology Working Group: Featured Images for August 2009." http://www.psi.edu/pgwg/images/aug09image.html.

Stoker, C. R. and C. Emmart, eds. *Strategies for Mars: A Guide to Human Exploration*. Science and Technology Series, Vol. 86. San Diego, CA: Univelt—American Astronautical Society, 1996.

Taylor, F. W. *The Scientific Exploration of Mars*. Cambridge, UK: Cambridge University Press, 2010.

Tillman, J. E. "Mars: Temperature Overview." http://www-k12.atmos.washington.edu/ k12/resources/mars_data-information/temperature_overview.html.

Zubrin, R. *How to Live on Mars.* New York: Three Rivers Press, 2008.

———. *The Case for Mars.* New York: Free Press, 2011.

For many more beautiful pictures of the surface of Mars, visit: http://beautifulmars.tumblr.com/.

OUR MOON

Biesbroek, R. and G. Janin. *Ways to the Moon?* ESA Bulletin 103 (August 2000).

Eckart, P. *Lunar Base Handbook.* New York: McGraw-Hill, 1999.

"Field Testing for the Moon." http://www.nasa.gov/exploration/analogs/then-and-now.html.

Heiken, G. H., D. T. Vaniman, and B. M. French. *Lunar Sourcebook: A User's Guide to the Moon.* Cambridge: Cambridge University Press, 1991.

"181 Things to Do on the Moon." http://science.nasa.gov/science-news/science-at-nasa/2007/02feb_181/.

"The Smell of Moondust." http://science.nasa.gov/headlines/y2006/30jan_smellofmoondust.htm.

"Spacelog Apollo 11." http://apollo11.spacelog.org/page/03:04:57:00/.

ATMOSPHERES OFF EARTH

Brooks, C. G., J. M. Grimwood, and L. S. Swenson Jr. *Chariots for Apollo: A History of Manned Lunar Spacecraft.* Washington, DC: NASA SP-4205, 1979.

"The Case of the Electric Martian Dust Devils." http://www.nasa.gov/centers/goddard/news/topstory/2004/0420marsdust.html.

Goody, R. *Principles of Atmospheric Physics and Chemistry.* New York: Oxford University Press, 1995.

"Mars." http://humbabe.arc.nasa.gov/MGCM.html.

Marshall, J., C. Bratton, J. Kosmo, and R. Trevino. "Interaction of Space Suits with Windblown Soil: Preliminary Mars Wind Tunnel Tests." SETI Institute, MS 239–12, NASA Ames Research Center, Moffett Field, CA. http://www.lpi.usra.edu/meetings/LPSC99/pdf/1239.pdf.

"NASA Toxicology Group." http://www1.jsc.nasa.gov/toxicology/ and references cited therein.

Ogilvie, K. W. and M. A. Coplan. "Overall Properties of the Solar Wind and Solar Wind Instrumentation." *Review of Geophysics* 33 Suppl. (1995).

Petit, D. "The Smell of Space." http://spaceflight.nasa.gov/station/crew/exp6/spacechronicles4.html.

Thompson, R. D. *Atmospheric Processes and Systems.* London: Routledge, 1998.

Wayne, R. P. *Chemistry of Atmosphere.* 3rd ed. Oxford: Oxford University Press, 2000.

RADIATION IN SPACE

Akasofu, S.-I. and Y. Kamide, eds. *The Solar Wind and the Earth.* Boston: Kluwer, 1987.

Barth, Janet. "The Radiation Environment." 1999. http://radhome.gsfc.nasa.gov/radhome/papers/apl_922.pdf.

Benestad, R. E. *Solar Activity and Earth's Climate.* Chichester, UK: Praxis, 2002.

Carlowicz, M. J. and R. E. Lopez. *Storms from the Sun: The Emerging Science of Space Weather.* Washington, DC: Joseph Henry Press, 2002.

Casolino, M., V. Bidoli, A. Morselli, L. Narici, M. P. De Pascale, P. Picozza, E. Reali, R. Sparvoli, G. Mazzenga, M. Ricci, P. Spillantini, M. Boezio, V. Bonvicini, A. Vacchi, N. Zampa, G. Castellini, W. G. Sannita, P. Carlson, A. Galper, M. Korotkov, A. Popov, N. Vavilov, S. Avdeev, and C. Fuglesang. "Space Travel: Dual Origins of Light Flashes Seen in Space." *Nature* 422, no. 680 (2003).

Catling, D. C., C. S. Cockell, and C. P. McKay. "Ultraviolet Radiation on the Surface of Mars." http://mars.jpl.nasa.gov/mgs/sci/fifthconf99/6128.pdf.

The Committee on Solar and Space Physics and the Committee on Solar-Terrestrial Research. *Radiation and the International Space Station: Recommendations to Reduce Risk.* National Research Council, 2000.

Cucinotta, F. A., F. K. Manuel, J. Jones, G. Iszard, J. Murrey, B. Djojonegro, and M. Wear. "Space Radiation and Cataracts." *Radiation Research* 156 (5 Pt 1) (November 2001): 460–466.

Dooling, D. "Digging in and Taking Cover: Lunar and Martian Dirt Could Provide Radiation Shielding for Crews on Future Missions." http://science.nasa.gov/newhome/headlines/msad20jul98_1.htm.

English, R. A., R. E. Benson, V. Bailey, and C. M. Barnes. "Average Radiation Doses of the Flight Crews for the Apollo Missions." In *Apollo Experience Report—Protection Against Radiation.* Houston: Manned Spacecraft Center, 1973.

Johnston, R. S., L. F. Dietlein, and C. A. Berry. *Biomedical Results of Apollo.* NASA SP-368, 1975.

Lilenstein, J. and J. Bornarel. *Space Weather, Environment, and Societies.* New York: Springer, 2006.

Miller, K. "The Phantom Torso." http://science.nasa.gov/headlines/y2001/ast04may_1.htm.

Miller, R. C., S. G. Martin, W. R. Hanson, S. A. Marino, and E. J. Hall. "Heavy-Ion Induced Oncogenic Transformation." *Center for Radiological Research Reports* 1998: 21–24.

"NASA Facts: Understanding Space Radiation." October 2002. FS-2002–10–080-JSC.

"The Natural Space Radiation Hazard." http://radhome.gsfc.nasa.gov/radhome/Nat_Space_Rad_Haz.htm.

Ohnishi, T., A. Takahashi, and K. Ohnishi. "Biological Effects of Space Radiation." *Biological Science in Space* Suppl: S203–10 (October 15, 2001).

Parker, E. N. "Shielding Space Travelers." *Scientific American* (March 2006).

"Radiation and Long-Term Space Flight." http://www.nsbri.org/Radiation/ (This site contains many useful links, only a few of which are cited here explicitly.)

Ramaty, R., N. Mandzhavidze, and X-M Hua, eds. *High-Energy Solar Physics*. Woodbury, NY: American Institute of Physics Press, 1996.

Saganti, P. B., F. A. Cucinotta, J. W. Wilson, and W. Schimmerling. "Visualization of Particle Flux in the Human Body on the Surface of Mars." http://www.ncbi.nlm.nih.gov/pubmed/12793743.

Saganti, P. B., F. A. Cucinotta, J. W. Wilson, L. C. Simonsen, and C. Zeitlin. "Radiation Climate Map for Analyzing Risks to Astronauts on the Mars Surface from Galactic Cosmic Rays." http://link.springer.com/article/10.1023%2FB%3ASPAC.0000021010.20082.1a.

Simonsen, L. C. and J. E. Nealy. "Mars Surface Radiation Exposure for Solar Maximum Conditions and 1989 Solar Proton Events." NASA Technical Paper 3300, February 1993.

"Single Event Effect Criticality Analysis." Section 3. 1996. http://radhome.gsfc.nasa.gov/radhome/papers/seecai.htm.

"Solar Iradiance [*sic*]." http://hyperphysics.phy-astr.gsu.edu/hbase/vision/solirrad.html.

"Space Station Radiation Shields 'Disappointing.'" http://www.newscientist.com/article.ns?id=dn2956.

Stern, D. P. and M. Peredo. "The Tail of the Magnetosphere." http://www-istp.gsfc.nasa.gov/Education/wtail.html.

Stewart, R. D. "The Nature of a Fatal DNA Lesion." Pacific Northwest National Laboratory-SA-30810, June 25, 2001.

Stozhkov, Y. I. "The Role of Cosmic Rays in the Atmospheric Processes." *Journal of Physics: Nuclear and Particle Physics* 29 (2003): 913–923.

Task Group on the Biological Effects of Space Radiation, Space Studies Board, and Commission on Physical Science, Mathematics, and Applications, National Research Council. *Radiation Hazards to Crews of Interplanetary Missions*. Washington, DC: National Academy Press, 1996. http://www.nap.edu/books/0309056985/html/R1.html.

Tobias, C. A. and P. Todd, eds. *Space Radiation Biology and Related Topics*. New York: Academic Press, 1974.

Townsend, L. W. "Overview of Active Methods for Shielding Spacecraft from Energetic Space Radiation." *Physica Medica* 17, Suppl. 1 (2001): 84–85.

"Understanding Space Radiation." NASA Fact Sheet FS-2002–10–080-JSC. October 2002. http://spaceflight.nasa.gov/spacenews/factsheets/pdfs/radiation.pdf.

U.S. Army Corps of Engineers. *Engineering and Design—Guidance for Low-Level Radioactive Waste (LLRW) and Mixed Waste (MW) Treatment and Handling*. EM 1110-1-4002, 1997.

"What Is Space Radiation?" http://srag-nt.jsc.nasa.gov/spaceradiation/what/what.cfm.

Wilson, J. W., F. A. Cucinotta, M-H.Y. Kim, and W. Schimmerling. "Optimized Shielding for Space Radiation Protection." *Physical Medica* XVII, Supplement 1 (2001).

Wilson, J. W., F. A. Cucinotta, H. Tai, L. C. Simonsen, J. L. Shinn, S. A. Thibeault, and M. Y. Kim. "Galactic and Solar Cosmic Ray Shielding in Deep Space." NASA Technical Paper 3682, 1997.

Wilson, J. W., J. Miller, A. Konradi, and F. A. Cucinotta, eds. *Shielding Strategies for Human Space Exploration*. NASA Conference Publication 3360, Dec. 1997. http://www -do.fnal.gov/~diehl/Public/snap/meetings/NASA-97-cp3360.pdf.

Wilson, J. W., J. L. Shinn, R. C. Singleterry, H. Tai, S. A. Thibeault, L. C. Simonsen, F. A. Cucinotta, and J. Miller. *Improved Spacecraft Materials for Radiation Shielding*, 2000. http://ntrs.nasa.gov/archive/nasa/casi.ntrs.nasa.gov/19990040361.pdf

SURFACE ACTIVITY ON WORLDS NEAR EARTH

"*Apollo 15* Soil Mechanics Investigation." http://www.lpi.usra.edu/expmoon/Apollo15/ A15_Experiments_SMI.html.

"*Apollo 17* Soil Mechanics Investigation." http://www.lpi.usra.edu/expmoon/Apollo17/ A17_Experiments_SMI.html.

Bell, E. T. "Crackling Planets." http://science.nasa.gov/headlines/y2005/10aug_crackling .htm.

Bottino, G., M. Chiarle, A. Joly, and G. Mortara. "Modeling Rock Avalanches and Their Relation to Permafrost Degradation in Glacial Environments." *Permafrost and Periglacial Processes* 13 (2002): 283–288.

Camp, Vic. "How Volcanoes Work." http://www.geology.sdsu.edu/how_volcanoes_work/.

Carr, M. H., W. A. Baum, K. R. Blasius, G. A. Briggs, J. A. Cutts, T. C. Duxbury, R. Greeley, J. Guest, H. Masursky, B. A. Smith, L. A. Soderblom, J. Veverka, and J. B. Wellman. "Craters." NASA SP-441: Viking Orbiter Views of Mars. (Many other useful and interesting images in SP-441 http://history.nasa.gov/SP-441/contents.htm.)

Caruso, P. A. "Seismic Triggering Mechanisms on Large-Scale Landslides, Valles Marineris." 34th Lunar and Planetary Science Conference, League City, TX, March 2003. Abstract 1525. http://www.lpi.usra.edu/meetings/lpsc2003/pdf/1525.pdf.

Committee on Planetary and Lunar Exploration. *Assessment of Mars Science and Mission Priorities*, Chapter 2. Washington, DC: National Academies Press, 2003. http:// www.nap.edu/books/0309089174/html/.

"Distribution Fan Near Holden Crater: PIA04869." *NASA Planetary Photojournal*. http://photojournal.jpl.nasa.gov/catalog/PIA04869.

Francis, P. *Volcanoes: A Planetary Perspective*. New York: Oxford University Press, 1993.

Harrison, K. H. and R. E. Grimm. "Rheological Constraints on Martian Landslides." *Icarus* 163, no. 2 (June 2003): 347–362.

Kious, W. J. and R. I. Tilling. *This Dynamic Earth: The Story of Plate Tectonics*. Online edition. http://pubs.usgs.gov/gip/dynamic/dynamic.html.

Krakauer, J. *Into Thin Air*. New York: Anchor Books, 1997.

"Lava Flows and Their Effects." USGS Volcano Hazards Program. http://volcanoes.usgs .gov/Hazards/What/Lava/lavaflow.html.

"Layers, Landslides, and Sand Dunes in Mars *Odyssey* Mission." http://themis.asu.edu/zoom-20031027a.

Lewis, John S. *Mining the Sky.* New York: Helix Books, 1996.

"Mars Global Surveyor." http://www.msss.com/moc_gallery/.

"Mars Orbiter Sees Landslide." http://spaceflightnow.com/news/n0111/05marslandslide/.

Marshall, J., C. Braton, J. Kosmo, and R. Trevino. "Interaction of Space Suits with Windblown Soil: Preliminary Mars Wind Tunnel Tests." SETI Institute, MS 239–12, NASA Ames Research Center, Moffett Field, CA. http://www.lpi.usra.edu/meetings/LPSC99/pdf/1239.pdf.

"Plate Tectonics on Mars?" http://science.nasa.gov/newhome/headlines/ast29apr99_1.htm.

"Recent Movements: New Landslides in Less than 1 Martian Year." MGS MOC Release No. MOC2–221, 12 March 2000. http://www.msss.com/mars_images/moc/lpsc2000/3_00_massmovement/.

Savage, D., G. Webster, and M. Nickel. "Mars May Be Emerging from an Ice Age." NASA press release 03–415. http://www.nasa.gov/home/hqnews/2003/dec/HQ_03415_ice_age.html.

Task Group on Issues in Sample Return. *Mars Sample Return: Issues and Recommendations.* Washington, DC: National Academies Press, 1997. http://www.nap.edu/catalog/5563.html#toc.

"Volcanology on Mars." https://en.wikipedia.org/wiki/Volcanology_of_Mars.

Wilson, L. "The Influence of Planetary Environments on Volcanic Eruption and Intrusion Processes." Paper presented at Planetary Geophysics Meeting, London, February 13–14, 2003. http://bullard.esc.cam.ac.uk/~nimmo/wilson.html.

Woods, A. W. "How They Explode: The Dynamics of Volcanic Eruptions." In *Annual Editions Geology 99/0,* ed. Douglas B. Sherman. Guilford, CT: Dushkin/McGraw-Hill, 1999.

WATER IN SPACE

Carr, Michael H. *Water on Mars.* New York: Oxford University Press, 1996.

"Clementine Bistatic Radar Experiment." National Space Science Data Center ID: 1994–004A-9. http://nssdc.gsfc.nasa.gov/nmc/experimentDisplay.do?id=1994–004A-09.

Cowen, R. "Taste of a Comet: Spacecraft Samples and View Wild 2." http://www.sciencenews.org/articles/20040110/fob1.asp.

"Distributory Fan Near Holden Crater." *NASA Planetary Photojournal.* http://photojournal.jpl.nasa.gov/catalog/PIA04869.

Head, J. W. and D. R. Marchant. "Cold-Based Mountain Glaciers on Mars: Western Arsia Mons." http://www.planetary.brown.edu/planetary/documents/2837.pdf.

———. "Mountain Glaciers on Mars?" Vernadsky Institute Microsymposium 36, October 14–16, 2002, Moscow, Russia.

Malin, M. C. and K. S. Edgett. "Evidence for Recent Groundwater Seepage and Surface Runoff on Mars." *Science* 288 (June 30, 2000): 2330–2335. http://www.sciencemag .org/cgi/content/abstract/288/5475/2330?rbfvrToken=e147c6ee6fff5c42735ec91da3f8 f3b8f21da4b4.

Morton, O. "Mars: Is There Life in the Ancient Ice?" *National Geographic* (January 2004).

Purves, W. K., G. H. Orians, H. C. Heller, and D. Sadava. *Life: The Science of Biology*. 5th ed. New York: W. H. Freeman, 1997.

"Sea Level and Climate." http://pubs.usgs.gov/fs/fs2–oo/.

"Water at Martian South Pole." http://www.esa.int/SPECIALS/Mars_Express/SEMY KEX5WRD_o.html.

IMPACTS IN SPACE

Aceti, R., G. Drolshagen, J. A. M. McDonnell, and T. Stevenson. "Micrometeoroids and Space Debris—The Eureca Post-Flight Analysis." *European Space Agency Bulletin* 80 (November 1994).

Bellot-Rubio, L. R., J. L. Ortiz, and P. V. Sada. "Observations and Interpretation of Meteoroid Impact Flashes on the Moon." In *Earth, Moon, and Planets*. Netherlands: Kluwer, 2002, 82–83, 575–598.

Cudnik, B. M., D. W. Dunham, D. M. Palmer, A. C. Cook, R. J. Venable, and P. S. Gural. "Ground-Based Observations of High-Velocity Impacts on the Moon's Surface—The Lunar Leonid Phenomena of 1999 and 2001, 2002." Paper presented at 33rd Lunar and Planetary Science Conference, Houston, TX, March 11, 2002. http://www.lpi .usra.edu/meetings/lpsc2002/pdf/1329.pdf

Gehrels, Tom, ed. *Hazards Due to Comets and Asteroids*. Tucson and London: University of Arizona Press, 1994.

Graham, G. A., A. T. Kearsley, I. P. Wright, M. M. Grady, G. Drolshagen, N. McBride, S. F. Green, M. J. Burchell, H. Yano, and R. Elliot. *Analysis of Impact Residues on Spacecraft: Possibilities and Problems*, Session 9: Microparticles. 3rd European Conference on Space Debris, Darmstadt, Germany; Noordwijk, Netherlands. ESA Publications Division, 2001.

Johnson, N. L. "Monitoring and Controlling Debris in Space." *Scientific American* (August 1998).

Kerridge, J. F. and M. Shapley-Matthews, eds. *Meteorites and the Early Solar System*. Tucson: University of Arizona Press, 1988.

Kiefer, W. S. "*Apollo 15* Passive Seismic Experiment." www.lpi.usra.edu/expmoon/ Apollo15/A15_Experiments_PSE.html.

"Leonid Flashers—Meteoroid Impacts on the Moon." http://iota.jhuapl.edu/lunar _leonid/beech99.htm.

"Leonids on the Moon." http://science.nasa.gov/newhome/headlines/ast03nov99_1 .htm.

Marshall, J., C. Bratton, J. Kosmo, and R. Trevino. "Interactions of Space Suits with Windblown Soil: Preliminary Mars Wind Tunnel Results." In *Studies of Mineralogical and Textural Properties of Martian Soil: An Exobiological Perspective Conference 1999*, 79.

Mehrolz, D., L. Leushacke, W. Flury, R. Jehn, H. Klinkrad, and M. Landgraf. "Detecting, Tracking, and Imaging Space Debris." *European Space Agency Bulletin* 109 (February 2002): 128–134.

Melosh, H. J. "Can Impacts Induce Volcanic Eruptions?" http://www.lpi.usra.edu/meetings/impact2000/pdf/3144.pdf.

"Micrometeoroids and Space Debris." NASA Quest. http://quest.arc.nasa.gov/space/teachers/suited/9d2micro.html.

Oritz, J. L., P. V. Sada, L. R. Bellot Rubio, F. J. Aceituno, J. Aceituno Gutiérrez, and U. Thiele. "Optical Detection of Meteoroidal Impact on the Moon." *Nature* 405 (2000): 921–923.

Rubio, L. R. B., J. L. Ortiz, and P. V. Sada. "Observations and Interpretation of Meteoroid Impact Flashes on the Moon." In *Earth, Moon, and Planets*. The Netherlands: Kluwer, 2000, 82–83, 575–598.

"*Salyut 7/Kosmos 1686*: Helium Tank." http://fernlea.tripod.com/tank.html.

Sumners, C. and C. Allen. *Cosmic Pinball: The Science of Comets, Meteors, and Asteroids.* New York: McGraw-Hill, 2000.

Technical Report on Space Debris. New York: United Nations, 1999. http://www.unoosa.org/pdf/reports/ac105/AC105_720E.pdf.

MEDICINE IN SPACE

Antunano, M. J. "Commercial Human Space Flight Medical Issues." February 2014. https://www.google.com/url?sa=t&rct=j&q=&esrc=s&source=web&cd=1&ved=0CCEQFjAAahUKEwiS2oGinPzHAhUFGR4KHfWHBMM&url=http%3A%2F%2Fwww.faa.gov%2Fabout%2Foffice_org%2Fheadquarters_offices%2Fast%2F17th_cst_presentations%2Fmedia%2FCommercial_Human_Spaceflight_Medical_Issues_Dr_Melchor_Antunano.pdf&usg=AFQjCNH4UsDVfjuxLDsezi9EBZaGU0oInA.

Asashima, M. and G. M. Malacinski, eds. *Fundamentals of Space Biology.* Tokyo: Scientific Societies Press and Springer-Verlag, 1990.

"Astronaut Blaha Says His Body Healed More Slowly During 118 Days on *Mir.*" *The Virginia-Pilot* (February 16, 1997).

Barry, P. L. and T. Phillips. "Mixed Up in Space." Science@NASA. http://science.nasa.gov/headlines/y2001/ast07aug_1.htm.

Beasley, D. and W. Jeffs. "Space Station Research Yields New Information About Bone Loss." http://www.nasa.gov/home/hqnews/2004/mar/HQ_04084_station_bone_loss.html.

"The Body in Space." http://www.nsbri.org/DISCOVERIES-FOR-SPACE-and-EARTH/ The-Body-in-Space/.

Bogomolov, V. V., et al. "International Space Station Medical Standards and Certification for Space Flight Participants." *Aviation, Space and Environmental Medicine* 78, no. 12 (December 2007).

Brown, A. S. "Pumping Iron in Microgravity." NASA Exploration Systems. http://www .nasa.gov/audience/forstudents/postsecondary/features/F_Pumping_Iron_in _Microgravity.html.

Buckley, J. C. Jr. *Space Physiology*. Oxford: Oxford University Press, 2006.

Buckley, J. C. and J. L. Homick, eds. *The Neurolab Spacelab Mission: Neuroscience Research in Space*. Washington, DC: Government Printing Office, NASA SP-2003535, 2003.

Cheatham, M. L. "Advanced Trauma Life Support for the Injured Astronaut." 3rd ed. http://www.surgicalcriticalcare.net/Resources/ATLS_astronaut.pdf.

Clement, G. *Fundamentals of Space Medicine*. 2nd ed. New York: Springer, 2011.

Committee on Space Biology and Medicine (Space Studies Board) and Commission on Physical Sciences, Mathematics, and Applications (National Research Council). *A Strategy for Research in Space Biology and Medicine in the New Century*. Washington, DC: National Academy Press, 1998.

Currier, P. "A Baby Born in Space." NASA Quest. http://quest.arc.nasa.gov/people/ journals/space/currier/08-26-99.html.

Czarnik, T. R. "Medical Emergencies in Space." http://chapters.marssociety.org/usa/oh/ aero5.htm.

"Dental Issues in Space." http://www.dental-tribune.com/mobarticles/content/scope/ news/region/americas/id/12838?mobsw=mob.

"The Disadvantageous Physiological Effects of Spaceflight." http://www.descsite.nl/ Students/DeHon/DeHon_chapter2.htm.

"Effects of Spaceflight on the Human Body." https://en.wikipedia.org/wiki/ Effect_of_spaceflight_on_the_human_body.

Farahani, R. M. and L. A. DiPietro. "Microgravity and the Implications for Wound Healing." *International Wound Journal* 5, no. 4 (2008).

Flight Crew Medical Standards and Spaceflight Participant Medical Acceptance Guidelines for Commerical Space Flight. Center of Excellence for Commercial Space Transportation, June 30, 2012. http://www.coe-cst.org/core/scripts/wysiwyg/kcfinder/upload/ files/2012.08.06%20Task%20183-UTMB%20Final%20Report.pdf.

Graveline, D. "Body Fluid Changes in Space." http://www.spacedoc.net/body_fluid .html.

Grenon, S. M., J. Saary, G. Gray, J. M. Vanderploeg, and M. Hughes-Fulford. "Can I Take a Space Flight? Considerations for Doctors." *British Medical Journal* 8, no. 124 (December 13, 2012).

Hall, T. W. "Adverse Effects of Weightlessness." http://permanent.com/s-nograv.htm.

"How Does Spending Prolonged Time in Microgravity Affect the Bodies of Astronauts?" http://www.scientificamerican.com/article/how-does-spending-prolong/.

Hullander, D. and P. L. Barry. "Space Bones." http://science.nasa.gov/headlines/y2001/ast01oct_1.htm.

International Space Station Environmental Control and Life Support System. NASA FS-2002–05–85-msfc, May 2002.

Kirkpatrick, A. W., M. R. Campbell, O. Novinkov, I. Goncharov, and I. Kovachevich. "Blunt Care and Operative Care in Microgravity." *Journal of the American College of Surgeons* 184, no. 5 (May 1997): 441–453.

Lu, Ed. "Expedition 7: Working Out." http://spaceflight.nasa.gov/station/crew/exp7/lu letters/lu_letter7.html.

McDonald, P. V., J. M. Vanderploeg, K. Smart, and D. Hamilton. "AST Commercial Human Space Flight Participant Biomedical Data Collection." Wyle Laboratories, Technical Report LS-09–2006–001, February 1, 2007.

Miller, K. "Space Medicine." http://science.nasa.gov/headlines/y2002/30sept_space medicine.htm.

Mitari, G. "Space Tourism and Space Medicine." *Journal of Space Technology and Science* 9 (1993). http://www.spacefuture.com/archive/space_tourism_and_space_medicine .shtml.

Modak, S., A. Krishnamurthy, and M. M. Dogra. "Human Centrifuge in Aero Medical Evaluations." *Indian Journal of Aerospace Medicine* 47, no. 2 (2003). http://www .medind.nic.in/iab/t03/i2/iabt03i2p6.pdf.

Moore, D., P. Bie, and H. Oser, eds. *Biological and Medical Research in Space.* Berlin: Springer, 1996.

Nave, C. R. "Cooling of the Human Body." http://hyperphysics.phy-astr.gsu.edu/hbase/ thermo/coobod.html.

O'Rangers, E. A. "Space Medicine." http://www.nss.org/community/med/home.html.

"Recommended Practices for Human Space Flight Occupant Safety." FAA, 2014. https:// www.google.com/url?sa=t&rct=j&q=&esrc=s&source=web&cd=2&ved=0C CgQFjABahUKEwiS20GinPzHAhUFGR4KHfWHBMM&url=https%3A%2F%2 Fwww.faa.gov%2Fabout%2Foffice_org%2Fheadquarters_offices%2Fast%2Fmedia %2FRecommended_Practices_for_HSF_Occupant_Safety-Version_1-TC14–0037 .pdf&usg=AFQjCNG_x_0Zn5-woxUsiwAq6b8USVFOhQ.

"Space Medicine." Japanese Aerospace Exploration Agency. http://iss.sfo.jaxa.jp/med/ index_e.html.

"Space Travel Increases Some Health Risks." Science@NASA. http://science.nasa.gov/ newhome/headlines/msad04nov98_1.htm.

"Study Suggests Spaceflight May Decrease Human Immunity." http://www.nasa.gov/ home/hqnews/2004/sep/HQ_04320_immunity.html.

"Sustained Acceleration." *United States Naval Flight Surgeon's Manual,* 3rd ed. Naval Aerospace Medical Institute, 1991, chapter 2.

Tobias, C. A. and P. Todd, eds. *Space Radiation Biology and Related Topics.* New York: Academic Press, 1997.

SOCIAL INTERACTIONS, MENTAL HEALTH, AND OTHER HUMAN FACTORS ISSUES

"Adult ADHD." http://www.webmd.com/add-adhd/guide/10-symptoms-adult-adhd.

Ball, J. R. and C. H. Evans Jr., eds. *Safe Passage: Astronaut Care for Exploration Missions.* Washington, DC: National Academy Press, 2001. http://www.nap.edu/books/0309075858/html/R1.html.

Bruno, F. J. "An Introduction to Symptoms of Boredom." http://www.thehealthcenter.info/emotions/boredom/.

Burrough, B. *Dragonfly: NASA and the Crisis Aboard MIR.* New York: HarperCollins, 1998.

Cataletto, A. E. and G. Hertz. "Sleeplessness and Circadian Rhythm Disorder." http://www.emedicine.com/neuro/topic655.htm.

"Claustrophobia."   http://www.betterhealth.vic.gov.au/bhcv2/bhcarticles.nsf/pages/Claustrophobia?OpenDocument.

Connors, M. M., A. A. Harrison, and F. R. Akins. *Living Aloft: Human Requirements of Extended Spaceflight.* Washington, DC: Government Printing Office, 1985.

Cooper, Jr., H. S. F. "The Loneliness of the Long-Duration Astronaut." *Air and Space Magazine* 2 (June–July 1996): 37–45.

Cowing, K. "It's Noisy Out in Space." http://www.spaceref.com/news/viewnews.html?id=831.

Cromie, W. J. "Astronauts Explore the Role of Dreaming in Space." *Harvard University Gazette,* February 6, 1997.

Czarnik, T. R. "Medical Emergencies in Space." http://chapters.marssociety.org/usa/oh/aero5.htm.

Dawson, S. J. "Human Factors in Mars Research: An Overview." In *Proceedings of the 2nd Australian Mars Exploration Conference, 2002,* ed. Jonathan D.A. Clarke, Guy M. Murphy, and Michael D. West. Sydney, Australia: Mars Society Australia, 2002.

Devitt, T. "High Living." http://whyfiles.org/124space_station/4.html. (All sections of the site are useful.)

Dingfelder, S. F. "Mental Preparation for Mars." *Monitor on Psychology* 35, no. 7 (July/August 2004).

Dudley-Rowley, M., S. Whitney, S. Bishop, B. Caldwell, and P. D. Nolan. "Crew Size, Composition, and Time: Implications for Habitat and Workplace Design in Extreme Environments." SAE 2001–01–2139. Paper presented at 31st International Conference on Environmental Systems, Orlando, FL, July 2001.

Dunn, M. "Serenity Is Scarce in Orbit." http://www.penceland.com/No_Serenity.html.

Epstein, R. "Buzz Aldrin, Down to Earth." *Psychology Today* (May/June 2001).

"Etiology of Anxiety Disorders." *Mental Health: A Report of the Surgeon General.* http://www.surgeongeneral.gov/library/mentalhealth/chapter4/sec2_1.html.

"Facts About Post-Traumatic Stress Disorder." National Institute of Mental Health Publ. No. OM-99–4157 (Revised) (Sept. 1999).

Harrison, A. A. *Spacefaring: The Human Dimension*. Berkeley: University of California Press, 2001.

Harrison, A .A., Y. A. Clearwater, and C. P. McKay. *From Antarctica to Outer Space: Life in Isolation and Confinement*. New York: Springer-Verlag, 1991.

"Hearing Lost in Space." http://news.bbc.co.uk/1/hi/special_report/iss/319323.stm.

Hutchinson, K. "Terminating T3." *The Antarctic Sun*, November 3, 2002.

"Internet Mental Health." http://www.mentalhealth.com/. (Many useful mental health sections.)

Lawson, B. D. and A. M. Mead. "The Sopite Syndrome Revisited: Drowsiness and Mood Changes During Real or Apparent Motion." *Acta Astronaut* 43 (3–6) (August–September 1998): 181–192.

Linenger, J. M.,*Off the Planet: Surviving Five Perilous Months Aboard the Space Station Mir*. New York: McGraw-Hill, 2000.

Long, J. *Mountains of Madness: A Scientist's Odyssey in Antarctica*. Washington, DC: Joseph Henry Press, 2001.

Long, P. W. "General Anxiety Disorder." http://www.mentalhealth.com/dis/p20-an07.html.

Morphew, M. E. "Psychological and Human Factors in Long Duration Spaceflight." *McGill Journal of Medicine* 6 (2001): 74–80.

Mundell, I. "Stop the Rocket I Want to Get Off." *New Scientist* 1869 (April 17, 1993).

National Institute of Mental Health. *Anxiety Disorders*. NIH Publication No. 02–3879. Washington, DC: National Institutes of Health, 2002.

The Numbers Count: Mental Disorders in America NIH Publication No. 01–4584.

"Ops-Alaska Space Exploration Human Factors." http://ops-alaska.com/.

"The Psychological and Social Effects of Isolation on Earth and Space." *QUEST—The History of Spaceflight Quarterly* 8, no. 2 (2000).

"PTSD." https://www.nlm.nih.gov/medlineplus/magazine/issues/winter09/articles/winter09pg10–14.html.

Roback, H. B. "Adverse Outcomes of Group Psychotherapy." *Journal of Psychotherapy Practice and Research* 9, no. 3 (summer 2000). http://www.ncbi.nlm.nih.gov/pmc/articles/PMC3330596/.

The Society for Human Performance in Extreme Environments. http://hpee.org/hpee.php.

Thomas, T. L., F. C. Garland, D. Mole, B. A. Cohen, T. M. Gudewicz, R. T. Spiro, and S. H. Zahm. "Health of U.S. Navy Submarine Crew During Periods of Isolation." *Aviation & Space Environmental Medicine* 74, no. 3 (March 2003): 260–265.

HUMAN-MADE CHALLENGES

Linenger, J. M. *Off the Planet: Surviving Five Perilous Months Aboard the Space Station Mir*. New York: McGraw-Hill, 2000.

POSTFLIGHT ADJUSTMENT

Epstein, R. "Buzz Aldrin, Down to Earth." *Psychology Today* (May/June 2001).

PROPULSION

Goebel, G. "Spaceflight Propulsion." http://www.vectorsite.net/tarokt.html.

Savage, D. "NASA Selects Teams to Lead Development of Next-Generation Ion Engine and Advanced Technology." NASA Press Release 02–118, June 27, 2002.

"Space Accidents." http://www.spacesafetymagazine.com/space-disasters/.

HISTORY

Burrows, W. E. *This New Ocean*. New York: Random House, 1998.

Chaikin, A. *A Man on the Moon*. London: Penguin, 2007.

"History: A Chronology of Mars Exploration." http://www.hq.nasa.gov/office/pao/History/marschro.htm.

"History of Lunar Impacts." http://iota.jhuapl.edu/lunar_leonid/histr224.htm.

Lauinius, R. and C. Fries. "Chronology of Defining Events in NASA History 1958–2003." http://www.hq.nasa.gov/office/pao/History/Defining-chron.htm.

"Moon and Planets Exploration Timeline." http://www.spacetoday.org/History/ExplorationTimeline.html.

Verne, J. *From the Earth to the Moon*. New York: Bantam Classics, 1993.

COMMERCIAL SPACE FLIGHT

Center of Excellence—Commerical Space Transportation. http://www.coe-cst.org/publications.html.

"IN FOCUS: How Space Tourists Are Prepared for Suborbital Flight." https://www.flightglobal.com/news/articles/in-focus-how-space-tourists-are-prepared-for-suborbital-364964/.

Space Safety Magazine. http://www.spacesafetymagazine.com/.

Space Station User's Guide. http://www.spaceref.com/iss/spacecraft/soyuz.tm.html.

MISCELLANEOUS

"Astronaut Training." https://en.wikipedia.org/wiki/Astronaut_training.

Bell, J. and J. Mitton, eds. *Asteroid Rendevous: NEAR Shoemaker's Adventures at Eros.* Cambridge: Cambridge University Press, 2002.

Fuller, J. "How Spacewalks Work." http://science.howstuffworks.com/spacewalk4.htm/printable.

ISS Medical Monitoring. ttp://www.nasa.gov/mission_pages/station/research/experiments/1025.html; http://www.ncbi.nlm.nih.gov/pubmed/18064923.

Mir. http://www.braeunig.us/space/specs/mir.htm.

Near-Earth Asteroids. http://airandspace.si.edu/exhibitions/exploring-the-planets/online/solar-system/asteroids/near.cfm.

Near Earth Object Program. http://neo.jpl.nasa.gov/news/news189.html; http://neo.jpl.nasa.gov/nhats/.

Plotner, T. *Moonwalk with Your Eyes: A Pocket Field Guide.* New York: Springer, 2010.

Principles Regarding Processes and Criteria for Selection, Assignment, Training and Certification of ISS (Expedition and Visiting) Crewmembers, Rev. A. http://www.spaceref.com/news/viewsr.html?pid=4578.

Space Debris. http://www.esa.int/Our_Activities/Operations/Space_Debris/Scanning_observing; http://orbitaldebris.jsc.nasa.gov/faqs.html; http://orbitaldebris.jsc.nasa.gov/protect/shielding.html;http://www.daviddarling.info/encyclopedia/W/Whipple_shield.html; https://the-moon.wikispaces.com/IAU+nomenclature.

Space Trainings. http://www.space-affairs.com/index.php?wohin=3rdfloor_p.

Young, L. R. "Artificial Gravity Considerations for a Mars Exploration Mission." *Annals of the New York Academy of Sciences* 871 (1999): 367–378.

INDEX

———

Page numbers in *italics* indicate figures and figure captions.

acceleration: extreme, 49–52; and *g*s, 27; into low Earth orbit, 151

accidents: and insurance, 46

adenosine triphosphate (ADT), 78

aerobic exercises: in space, 78–79

agoraphobia, 137

air friction, 24, 29

air pressure: on Earth, 103; on Mars, 103

albedo: on Moon, 179, *182*

alcohol: in space, 257

alcoholism, 137

Aldrin, Buzz (astronaut), *146*, 188–189

Alien (film), 25

Alone (Byrd), 118, 132

alpha particles: and cosmic rays, 90

alpha waves, 80

Ames Research Center (Moffett Field, CA), 159

ammonia, frozen, 12

Amor asteroids, 8, *11*, 35–36

Amor (1221) asteroid, *11*

angular momentum, conservation of: in microgravity, 154–155

Antarctica: and preflight training, 58; returning from, 249

Antarctic Circle: and biological clock and sleep disturbances, 80, 81

anxiety, 112, 123–124, 138

Apennine Mountains (Moon). *See* Montes Apenninus

Apollo asteroids, 8, 10, *11*, 35–36

Apollo (1862) asteroid, *11*

Apollo astronauts, 33, 161; food, 129; Moon rocks, 188; space suits, 165

Apollo 11, 13, 176, 188–189

Apollo landing sites, 177, *178, 179*

Apollo Lunar Surface Experiment Package (ALSEP), *178, 179*

Apollo missions, 33, 99, 151, 166, 169, 170, *171, 174, 175, 176, 181,* 188–189

Apollo spacecraft, 28

Arabia Terra (Mars), *209*

Arctic Circle: and biological clock, 80

Armstrong, Neil (astronaut), 188

Arsia Mons (Mars), *199*

ascent to orbit, 151

Ascres Mons (Mars), *199*

Asteroid Belt, 8, 9, 17

asteroids, vii, 7–12; composition of, 11–12, 192–193; impacts on, 105; journeys to, 18, 34–37, 189–190; in science fiction, 8–9

astronomy: from space, 157–158

Aten asteroids, 8, 10, *11*, 35–36

Aten (2062) asteroid, *11*

Atira asteroids, 8, *11*

Atira (16393) asteroid, *11*

atomic nuclei, 20; and cosmic rays, 90
atoms, 4, 16; electrical properties of, 19–20.
    See also electrons; neutrons; photons
attention-deficit hyperactivity disorder
    (ADHD), 135
aurorae, 90

Baikonur Cosmodrome, 3
balance: on return to Earth, 258
ballistic capture trajectory, 37
ballistic flight, 24
barycenter, 257
basalts: on Moon, 168
bathroom skills: in space, 58–59
beta waves, 80
B-52 motherships, 28
biological clock, 80, 81
bipolar disorder, 138, 139–140
Black Point Lava Flow (AZ): preflight
    training, 58
Blaha, John, 110, 111
blood pressure: in space, 70
Blue Origin, 41
body fluid redistribution, 70
body length: growth of in space, 73
bone activity: in space, 74, 256
bone marrow, 75
bone mineral loss, 75–76
bones: in microgravity, 75–77
booster rockets, 66, 67
boredom: in space, 132–141
brain terrain: on Mars, 218, 220
bread: in space, 153
Browning, Robert, 122
burn (rocket on) stage, 29; Near Earth
    Objects trips, 35
Byrd, Adm. Richard E., 118, 132

calcium: bone loss in space, 75
caldera, 211
Caloris basin impact (Mercury), 226
cameras: use in space, 145–146

cancer, 93, 138
Cape Canaveral (FL): accident, 46
captain (of space missions), 120–122
carbon atom, 19
carbon dioxide, frozen, 12. See also dry ice
carbon monoxide: frozen, 12
Carr, Gerald, 110
cataracts, 22, 93
catena, 180–181
Catena Davy (Moon), 183
caves: on Mars, 230; on Moon, 168, 183–184
CCD cameras, 145
CCDs (charge-coupled devices), 255
Center of Excellence for Commercial Space
    Transportation, 43
centrifuge: in preflight training, 51–52
Ceres, 7
chaos terrain: on Mars, 225–226, 227
chase planes: suborbital flights, 65
Chief Joseph, 248
children: and Mars colonization, 241,
    242–243, 244
chronic asthenia, 125–126
Churyumov, Klim Ivanovych, 16
circadian rhythm, 80–82; on long-haul
    space voyages, 192
circulatory system: in space, 70
claustrophobia, 126–127
clinical depression, 137
colors: in astronomy, 21
Columbia space shuttle: space debris
    impact, 97
comas, 15
Comet Halley, 16
Comet Hartley 2, 106
comet nuclei, 13, 14
Comet 67P/Churyumov-Gerasimenko, 16,
    17, 163, 193
comets, vii, 7, 12–16; composition of,
    192–193; day-night cycle, 81; impact
    flashes, 101; impacts on, 105; journeys to,
    18, 34–37, 189–190, 194

comet tails, 15
compression (of air), 25
Conrad, Charles, Jr. (astronaut), *179*
cosmic rays, 12, 89–92, 256; on Mars, 236; primary, 90–91; space photography and, 146–147
cosmic ray showers, *91*
Crater Stickney (Phobos), *39*
craters: on Mars, 212, *213–217,* 223; from micrometeorite impacts, 96–98; on Moon, 173–175, *176, 177;* pedestal, 210
Crater Theophilus (Moon), *175*
Cretaceous-Tertiary (K-T) extinction, 8
crowding: in spacecraft, 113–116
cultural differences: of people in space, 119–120
Cummings, Ray (writer), 4
Curiosity rover, 210
cyclothymia, 138
Cydonia (Mars), 203, 205

dark side of the Moon, 162
daydreaming, 133
day-night cycle, 80–81
deceleration: preflight training, 50, 51
decibels (db), 83
decompression: on space stations, 158
decompression sickness (the bends), 57
dehydration, 93
Deimos (moon), *39,* 193, 194–196; day-night cycle, 81; orbit of, 194–195; trips to, 38, 41
dementia, 92
depression, 125, 137, 139, 140; in space, 111, 137–140
descent from space: orbital flights, 160–161; suborbital flights, 149–150
despondency: in space, 137–140
deuterium, 19
Devil's Tower, 204
Devon Island, Canada: Mars simulation, 58
Dezhurov, Vladimir (cosmonaut), 141
diapers: ascent and return to Earth, 63

diarrhea, 93, 124
digestion: in space, 71–72
DNA damage, 22, 89
dorsa (on Moon), 178–179
dreaming: in space, 86–89
drinking: fluids in space, 72
drug absorption: in microgravity, 79
drug abuse, 137
drugs: in space, 257
dry ice: on Mars, 199, 207
dunes: on Mars, 217, *218, 219, 220*
dust devils and tracks: on Mars, 218–222
dust storms: on Mars, 228
dwarf planets, 6, 7
dysconjugate gaze (cross-eyed), 79
dyspnea, 88, 123
dysthymia, 138

Earth: atmosphere of, 5; equator width, 6; gravitational attraction on, 5; magnetic fields of, 30; readaptation to, 245–252
Earth-crossing asteroids (Near Earth Objects), 8, 9, 10, 34
Earth-Mars distance, 18–19
Earth-Moon distance, 18
Earth-Moon system: 24-hour light cycle, 81
Earth-orbit flights, 29–37
earthquakes, 99
Earth-Sun distance, 18
eating: in space, 71–72, 153–154
eclipses, 257
ecliptic, 7, 13, 34, 35
Einstein, Albert: special theory of relativity, 191
ejecta: from impacts, 173
ejecta blanket: on Moon, 173–174
electric charge: of protons and electrons, 19–20
electric vehicles, 167
electrolysis, 187
electromagnetic force, 20

electromagnetic radiation, 20–22, 23; and speed of light, 191

electrons, 4, 19–20

Empire Strikes Back, The (film), 8–9

Endurance crater (Mars), 217

erections: in space, 160

Eris, 7

Eros asteroid, 8, 9, 10, 81

erythyma, 93

European Space Agency, vii, 16

exoplanets, 255

extreme acceleration, 49–52

extreme anxiety, 123, 124

eye damage: from Sun, 157

Falcon 9 rocket, 46

fantasizing, 133

far side of the Moon, 162, 163; scarp on, 170; telescopes on, 186

fast-twitch (white) muscles, 78

fatigue, 93, 123, 124

Federal Aviation Administration, 43

feet: in microgravity, 73

femurs: bone mineral loss, 75

flameout, 28

flatulence: in space, 130

Flight Crew and Medical Standards and Spaceflight Participant Medical Acceptance Guidelines for Commercial Space Flight (2012), 44

Flight Indoctrination Program, 54

food: in space, 129–130

foot-drop pressure, 73

fossa: on Moon, 167

free fall, 27

fretted terrain: on Mars, 200

Gagarin, Yuri, 3

galactic cosmic rays. See cosmic rays

gamma rays, 22, 23

gamma waves, 80

gases: nature of, 5

Gemini astronauts: food, 129

Gemini spacecraft, 28

Gemini 12, 146

Gerasimenko, Svetlana Ivanovna, 16

Gibson, Edward, 110

Girl in the Golden Atom (Cummings), 4

glaciers: on Mars, 218

glaucoma: from space travel, 43

globular clusters of stars, 257

golf: on Moon, 166

gravitation, 20, 22–23

gravity, artificial, 256

gravity, law of: on Moon, 165–166

Great Dying. See Permian-Triassic extinction

green environment: on Mars, 237

grief: and long space voyages, 140–141

grooming: in space, 156

group dynamics: in space, 112–121

gs, 27, 49

gullies: on Mars, 223–224

Hadley, John, 169

Hadley Rille (Moon), 169, 174

hair: in space, 93, 156

Hall, Asaph, 195

Halley's comet, 16

hallucinations: in space, 127–128

Hartley 2 comet, 106

Haumea, 7, 13–14

Hawaiian Islands: preflight training, 58

healing processes: in space, 76

hearing loss, 83

heart attacks, 138

helium nuclei. See alpha particles

hematite: on Mars, 214–216, 217

Herschel, William, 21

Hertz, Heinrich, 21

Hess, Victor, 89

highlands: of Moon, 168

Hohmann transfer orbit, 34, 35, 37, 189–190

homesickness: and space travel, 135–136, 141, 243–244

hoodoos, 204

human growth hormone, 87

human spaceflight: early stages, 3; prices for, 3–4

humidity: in space, 132

hydrogen, liquid: as vehicle fuel on Moon, 167

hydrogen atom, 19

hypercalciuria, 76

hyperventilation, 123

hypothalamus: and sleep disturbances, 87

hypothyroidism, 138

hypoxia, 54–55

ice: defined, 12; on Moon, 186

ice caps: on Mars, 207

ileus, 71

impact craters, 99, 256; complex, 175; on Mars, 213; on Moon, 168, 173–175

impact flashes, 100–101

impacts: ejecta from, 173–174; in low Earth orbit, 94–99; and mountain building, 173; in space, 94–106; from wind, 104

impact study satellite, 98

impulsivity, 133, 135

incontinence, 137

informed consent form, 46

infrared radiation, 21, 23

inner ear, 52–53

Inner Earth Object asteroids, 35–36

insurance: for spaceflight, 44–47

Interior Earth Objects, 8

International Astronomical Union, 6; catalog of Moon features, 167

International Space Station, vii, 3, 6, 22, 29, 30, 31, *156*; air pressure in, 41; climate control, 131–132; and cosmic rays, 91; cupola of, *152, 153*; food, 129; long-term physical adjustments, 74; noise levels, 83; orbit of, *31*; sleeping compartments, *117*; smell malfunction, 130–131; trips to, 18; visitor standards, 43; Whipple shields, 98–99; Zvevda Space Module, *115*

Into Thin Air (Krakauer), 112

Irwin, James (astronaut), 169, 170

isolation, 136–137

isotopes, 19

Ivans, Marsha (astronaut), *156*

Jupiter, 6; Trojan asteroids, 10

Karman, Theodore von, 5

Karman line, 5–6, 30; orbital flights, 68; suborbital flights, 65

Kazakhstan, 3, 16

Kelly, Mark (astronaut), 74

Kelly, Scott (astronaut), 74, 76

Kevlar: Whipple shields, 98

kidney stones, 77

kinetic energy, 90

Korniyenko, Mikhail (cosmonaut), 74, 76

Krakauer, Jon, 112

Krieger crater (Moon), *181*

Kübler-Ross, Elisabeth, 140

Kuiper, Gerard, 13

Kuiper belt, 13–14

lacus: on Moon, 178

Lacus Mortis (Moon), *180*

Lagrange points, 11

landslides: on Moon, 175

launch, 63–67; suborbital flights, 64–65; to orbit, 150–151

lava: on Moon, 168, 169, 177

lava tubes: on Moon, 183

leadership: in space missions, 120–121

Lee scarp (Moon), 170

legal issues: spaceflight and, 44–47

Leonid meteor shower, *100*

Leonov, Alexi (Soviet cosmonaut), 56–57
liability, third-party, 46
lift-off phase, 66; orbital flights, 151; for suborbital flight, 26
light: speed of, 20, 191. *See also* visible light
lighting cycles: in space, 81
Lincoln scarp (Moon), 170
linear momentum: in microgravity, 155
Lineger, Dr. Jerry, 73, 244, 248, 250, 251
liquid hydrogen: as vehicle fuel on Moon, 167
liquids: nature of, 5
Long, John, 249–250
Long Duration Exposure Facility, 95, 98
long-period comets, 15–16
long-term physical adjustments to space, 74–106
low air pressure: hypoxia, 54–55
low Earth orbit, 30; day-night cycle, 81; launch to, 67
lump in throat, 88
Luna spacecraft, 162, 168
lunar insertion burn, 161, 162
lunar lander, 34, 255
lunar orbit, 162
lunar-orbiting space stations, 162
lunar phases, 164
lunar rovers, 169
lunar transient phenomena, 188–190
Lunokhod rover sites, 177

magma: in Moon, 168
magnetic fields: of Earth, 30; and space debris, 94–95; Van Allen belts, 32
major depression, 138
Makemake, 7, 255; orbit of, 13–14
mania: and bipolar disorder, 140
manic depression. *See* bipolar disorder
mare (pl. maria): on Moon, 167–168
Mare Ibrium (Moon), *177*
Mare Nubium, *183*
Mare Tranquilitatis (Moon), *185*

Mare Procellarum (Sea of Storms), 170
Mariner 7 spacecraft, 38
Mars, vii, 6; air pressure on, 103; atmosphere of, 198, 235; brain terrain on, 218, 220; canyon walls, 223, *224;* common surface features, 215–217; craters, 212, *213,* 223; day-night cycle, 81; dust and wind, 103–105; emigration to, 241–243; habitats, 196–197, 239; impacts on, 103–105; landing on, 39–40, 41, 196, 239; life on, 197; mountains on, *199;* permanent habitats, 229; planitia on, 172; polar regions, 207–208; radio signal time to Earth, 191; sedimentary layers, 207–208, *209,* 210; shorelines on, 214–215; sky color, 226–227; sunrise and sunset, 227; surface features to visit, 198–226; terraforming, 239; transportation on, 238; trips to, 18–19, 37–41, 196–230; Trojan asteroids, 10; ultraviolet radiation, 104–105; unknown features, 230; volcanoes on, 208–211; water ice clouds, 198; water on, 197, 201–203, 213, 223; winds and storms, 228
Mars colonization, 233–244: agriculture, 236–237; breathable air, 234–235; greenhouses, 237; radiation protection, 235–236; secondary habitat requirements, 237–239; social and mental health needs, 240–241; social constraints, 243; water, 234; work of settlers, 240–241
Mars Desert Research Stations, 108
Mars moons, 184–196; journeys to, 189–190. *See also* Deimos; Phobos
Mars rovers, 229
Martian, The (film), 37, 40, 103
Maxwell, James Clerk, 21
meals: in space, 255
medical history and testing: and eligibility for space travel, 42–44
Medusa Fossae (on Mars), 200–201, *202*
melanin, 22
melatonin, 80

mental health: space travel and, 108, 110–112. *See also* depression

Mercury, 6; albedo features, 180; chaos terrain, 225, *226*; travel to, 17

Mercury spacecraft, 28

meteorites, 94

meteoroids, 7, 12, 96; interplanetary space, 102; Moon strikes, 99

meteors, 101–102

meteor showers, *100*, 101–102

methane, frozen, 12

microgravity, *24*; acclimating to, 33; bones in, 75–77; brain's adaptation to, 68; docking with space stations, 151–152; drinking fluids in, 72; and drug absorption, 79; eating in, 71; medical adjustments to, 74–80; movement in, 154–156; preflight training, 52–54; proprioception, 69; and sleep, 87; and suborbital flight, 29; sense of smell in, 131; sex in, 149; spine and feet in, 73; tooth decay, 77; while orbiting Moon, 161. *See also* weightlessness

micrometeorites: impact craters, 96–98, 105

microwaves, 21, *23*

Mie, Gustav, 227

Mie scattering, 227

Millikan, Robert A., 89

Mir space station, 3, 250; depression on, 110, 111; noise, 83

Mission to Mars (film), 41

mixed mania, 140

montes (pl. mons): on Moon, 173

Montes Apenninus (Moon), *171, 174*

Montes Jura (Moon), *177*

Moon, vii; atmosphere absence, 164; basalts, 168; caves, 183–184; craters, 168, 173–175; day-night cycle, 81, 164; daytime and nighttime temperatures, 163; dust on, 164; far and near sides of, 162; geological features on, 167–182; highlands of, 168; ice on, 186; impacts on, 99–102; international landmarks, 177; landing experience, 162–163; landing sites and habitats, 163–164, 168, 177; meteor showers, 102; natural openings on, 182–*185*; regolith of, *13*, 173; scarps on, 170; sex on, 189; trips to, 3, 18, 32–34, 161–189; tunnels on, 182–184; vehicular travel on, 167; walking on, 165; water mining operations, 186–187

moonquakes, 99, 175

Moon rocks, 187–188

moons, 6, 7

morale: on space voyages, 132–141

Moses Lake, Washington State: preflight training, 58

motherships, 28; suborbital flights launch, 64–65

motion sickness, 88

motor skills: in orbital flights, 68–69

mountains. *See* montes

Mountains of Madness: A Scientist's Odyssey in Antarctica, 249–250

muscle loss, 74

muscles: and bones, 75; in space, 78–79

muscle tissue: atrophy in space, 78, 79

NASA (National Aeronautics and Space Administration), 255; Ames Research Center, 159; *Apollo 11* mission, 188–189; liftoff countdown, 66; Long Duration Exposure Facility, *95*; lunar seismometers, 99–100; Near Earth Objects catalog, 10; Office of Planetary Protection, 198; preflight training sites, 58

National Institute of Mental Health, 137

nausea, 93, 123; antinausea patch, 158

Naval Air Station, Pensacola, 54

Near Earth asteroids (NEAs), *11*

Near Earth Objects (Earth-crossing asteroids), 8, 9; NASA catalog of, 10; trips to, 34

NEAR Shoemaker lander, *10*

near side of the Moon, 162, *163*

Neptune, 6; Trojan asteroids and, 10

neutral buoyancy, 54
neutrons, 4, 19
Newton, Isaac, 21
Newton crater (Mars), *224*
noise: defined, 82–84; effects of, 83; in space, 84
noise-reduction technology: in space, 84
nucleus (of atom), 19
nutrition: in space, 71–72

obsessive-compulsive disorders, 137
Occam's razor, 203, 204, 257
Oceanus Procellarum (Ocean of Storms), 171
odors: in space, 130
Office of Planetary Protection, 198
Off the Planet (Linenger), 244, 248
Olympus Mons (Mars), *199*, 210–211
Oort, Jan, 14
Oort comet cloud, 14, 16
open clusters of stars, 257
Opportunity rover, *216*, *217*, 229
orbital flights, 150–151; launch, 65–67
orbits: and gravitation, 22
Orion spacecraft, 28
osteoporosis, 75–76, 77
otoliths, 52–53, 75
ozone: on Mars, 198

paludes: on Moon, 171
Palus Epidermirum (Marsh of Epidemics), 171
Palus Putredinis (Marsh of Decay), 171
Palus Somni (Marsh of Sleep), 171
panic attacks, 126, 137
parabolic flight, *24*, 28
parabolic trajectory: preflight training, 52
Parkinson's disease, 138
Pavonas Mons (Mars), *199*
pedestal craters, 210, *212*
pelvis: bone mineral loss, 75
Pensacola Naval Air Station, 54
permafrost: on Mars, 199

Permian-Triassic extinction (The Great Dying), 8
personality disorders, 137
personal space: in spacecraft, 113–115
Philae lander, 16
Phobos (moon), 39, 192–193, 194–196; day-night cycle, 81; orbit of, 194–195; trips to, 38, 41
photography: in space, 145–147
photons: properties of, 20–22; wavelengths of, 20–21
physical activity: and boredom, 134
pituitary gland: and sleep disturbances, 87
planets: defined, 6–7; orbits of and asteroids, 10–11
planitia, 172
Planitia Descensus (Moon), 172
plastics: smell of, 130
Pluto, 7; orbit of, 13–14
Pogue, William, 110
polar ice caps: on Mars, 199
post-traumatic stress disorder (PTSD), 124–126, 257
powers of ten, 253–254, 255
preflight training, 48–59
pregnancy: and space travel, 43; and Mars colonization, 242
primary cosmic rays, 90, 91
privacy, 116–118; and stress, 123
promontorium: on Moon, 175–176
promontorium Heraclides, 176, *177*
promontorium Laplace, 176, *177*
proprioception, 69, 165, 250
protons, 4, 19–20
Proxima Centauri, 16–17
psychological testing: for space travel, 43–44
psychosocial aspects of time in space, 49
pulling *g*s, 50–51, 54

radiation: and space medicine, 79; effect on humans, 92–93; exposure to in space, 89–93; fear of, 128

radiation phobia, 128
radio telescopes: on Moon, 186
radio waves, 21, *23*
rarefication (of air), 25
Rayleigh, Lord, 227
Rayleigh scattering, 227
recreation: on long-haul space voyages, 192
regolith: on Mars, 196; on Moon, 168, 175
regolith dust: on Moon, 163–164
regulations: conformity to, in space, 142
Reiner Gamma (Moon), 179, *182*
relativity, special theory of, 191
return to Earth, 244–252
rilles, *180, 181,* 182–183; on Moon, 168–169,
      182–183
rima. *See* rilles
Ritter, Johann Wilhelm, 21
riverbeds, dry: on Mars, 213, *214*
rockets: booster, 66, 67; liquid-propelled, 66
Romanenko, Yuri (cosmonaut), *77*
Röntgen, Wilhelm Conrad, 23
Rosetta spacecraft, *17,* 193
rupes. *See* scarps

Safe Passage: Astronaut Care for Exploration
      Missions, 111
Salyut series of space stations, 3
Salyut 6 space station, 77
Saturn, 6
scalloped topography: on Mars, 201, *202*
scarps: on Moon, 170
schizophrenia, 137
science fiction: asteroids in, 8–9; *Mission to*
      Mars (film), 41; *The Martian* (film), 103;
      sound in space, 25
Scott, David (astronaut), 166, 169, 170
screening processes: for space travel,
      109–110
seasonal affective disorder, 138
secondary cosmic rays, 90–91
seismometers: on Moon, 99–100
selfies: in space, 145, *146*

semimajor axis, *11*
senses: attacks on in space, 128–132
sex: in space, 159–160; in suborbital flights,
      148–149; on Moon, 189
sexual abuse, 138
Shephard, Alan (astronaut), 166
short-period comets, 16
showering: in space, 58–59
Simon, Alvah, 248–249
Simonyl, Charles, vii
simulators: in preflight training, 57–58
sinus: on Moon, 176–177
sinus Iridium (Moon), *177*
situational depression, 137
skin cancer, 22
skin care: in space, 156
Skylab, 3
Skylab 4 crew, 110
sleep disorders: in space, 86–89
sleep disturbances, 82–91; on return to
      Earth, 251
slope streaks: on Mars, 222, *223*
slow-twitch (red) muscles, 78, 79
small solar system bodies, 6, 7
smells: in spacecraft and space, 130–131
snoring: in space, 89
snow: on Mars, 198
snow blindness, 22
soccer: on Moon, 166
social interaction: of Mars settlers, 240
solar cells: on Moon, 167
solar eclipses: seen from the Moon, 164–165
Solar Max mission, *97*
solar power: on Mars, 238
solar wind, 12, 90
sound, 25; and noise, 83; in space, 84, 194
South Atlantic Anomaly, 31, *32*
Soviet Union: human spaceflight, 3; Moon
      trips, 32. *See also Mir* space station
Soviet-U.S. space race, 3
Soyuz-4 crew, 110
Soyuz space capsules, 28, 31

Soyuz T-14 mission, 110

space: defined, 5–6; impacts in, 94–106; sound in, 25

Space Adaptation Syndrome (space sickness), 53, 68, 151, 161, 256

space capsule: return to Earth, 161

spacecraft: meteoroid strikes, 96; and vibrations, 85

space debris, 94–106

Space Exploration Technologies Corporation (SpaceX), 41, 46

spaceflight insurance, 45

space law, 47

space medicine, 43, 79; broken bones and bone loss, 76, 77

space plane lander, 160

Spaceport America (AZ), 46

space shuttle: impacts, 96, *97*

space sickness. *See* Space Adaptation Syndrome

space stations: docking with, 151; Earth-orbiting, 3; lunar-orbiting, 162

space suits, 55–57; launch, 63–64; on Moon, 165; orbital flights launch, 65–66; trips to Moon, 162

space tourism, vii

space travel: options, 18; psychological and sociological aspects of, 107–142; training, 48–59

space trip preparation, 42–47

space voyages: communicating with Earth, 190–191; long-term, long distance, 190–230

space walking, 133, 158–159; and space suits, 56–57

SpaceX. *See* Space Exploration Technologies Corporation

spine: bone mineral loss, 75

sports: playing on Moon, 166

Star Wars franchise, 8

stereo cameras: use in space, 145–146

Stickney crater (Phobos), 195

Stickney-Hall, Angelina, 195

stress: on space voyages, 121–126; symptoms of, 122–123

strong force, 20

Styron, William: on depression, 138–139

subjective time, 133, 134–135

submarine missions, 111

suborbital flights, 18, 26–29, 147–150; launch, 64–65

Sun: eye damage from, 157

sunlight, 81

superchiasmatic nucleus, 80

sweating: and anxiety, 123

tablet computers: use in space, 145

tachycardia, 123

taste: in space, 129

tectonic plate movement: on Mars, 206

teeth: in space, 77–78

telescopes: on Moon, 185–186

temperature: comfort level in space, 131–132

territoriality: in space and spacecraft, 118–119

Thagard, Norman, 141

Tharsis Montes (Mars), *199*, 210

Theophilus crater (Moon), *175*

theta waves, 80

third-quarter phenomenon, 141

three-dimensional cameras, 145–146

Tiangong 1, 3

tidal pull, 195

tides: Phobos and Mars, 195

time: concept of by astronauts, 134–135

time cycles: in space, 80–81

Tito, Dennis, vii

trances: in space, 127–128

transient lunar phenomena, 188–190

translunar insertion burn, 161

translunar trajectory, 33

tritium, 19

Trojan asteroid (2010 TK), 34, 35, 36; communicating with Earth, 190

Trojan asteroids, 10–11; trips to, 18
tumors, 93
tunnels: on Mars, 182–184; on Moon, 168, 169
twenty-four hour sleep/wake cycle, 192
twinkling, 157, 164
2001: A Space Odyssey (film), 37
2016 TK. *See* Trojan asteroid

ultraviolet radiation, 21–22, 23; and space suits, 104–105
underground rivers: on Moon, 168, 169
United Nations: space accidents, 47
Uranus, 6; Trojan asteroids, 10
Usachev, Yury V. (cosmonaut), *117*
U.S. Navy: Flight Indoctrination Program, 54; submarine training, 111
Utopia Planitia (Mars), *202*
UV-A, UV-B, UV-C photons, 22

Valles Marineris, *199*, 205–206, 207, *227*; canyon wall, *225*
vallis: on Moon, 181–182
Vallis Alpes (Moon), *184*
Van Allen, James, 31
Van Allen radiation belts, 31, *32*, 34, 161; space debris, 94–95
Vasyutin, Vladimir (cosmonaut), 110
Venus, 6; travel to, 17
Verne, Jules (writer), 8
vertigo: orbital flights, 68
vibration frequencies: effects on human bodies, 85–86
vibrations: effects on activities, 87; and physical problems, 88; in space, 84–86
View-Master viewer, 146
Viking 2 spacecraft, 38
Villard, Paul, 22

virtual human software, 256–257
visible light, 20–21, 23
visual skills: in orbital flights, 68–69
Vitamin D, 22
volcanoes: on Mars, 208–211; on Moon, 173
Volynov, Boris (cosmonaut), 110
Vomit Comets, 53–54
vomiting, 93; and anxiety, 123; in space, 158; space sickness, 53
Voss, James S., *115*
Vostok-K rocket, 3

water: on Mars, 197, 201–203, 213, 223
water-mining operations: on Moon, 186–187
weak force, 20
weightlessness, 26, 27, 29; and personal space, 113–114; preflight training, 52–54; suborbital flights, 64, 147–148; trip to Mars, 37. *See also* microgravity
Wheatstone, Sir Charles, 146
Whipple shields, 98–99
white light, 21
white room, 65
William of Occam, 203
winter blues. *See* seasonal affective disorder
withdrawal: while in space, 136–137
work, 257; during space voyages, 159; of Mars settlers, 240–241
wrinkle ridges (dorsa), 178, 179, *181*

X-15 space planes, 28
X-rays, 22, 23

yardangs, 201

Zholobov, Vitaly (cosmonaut), 110
Zvevda Service Module, *115*